The Sacrifice

University of
Chester

New Directions in the Human-Animal Bond
Alan M. Beck, series editor

The Sacrifice

How Scientific Experiments Transform Animals and People

Lynda Birke,
Arnold Arluke, and
Mike Michael

Purdue University Press
West Lafayette, Indiana

ISBN 978-1-55753-432-3
1-55753-432-2

Library of Congress Cataloging-in-Publication Data
Birke, Lynda I. A.
The sacrifice : how scientific experiments transform animals and people /
Lynda Birke, Arnold Arluke, and Mike Michael.
    p. cm. -- (New directions in the human-animal bond)
Includes bibliographical references and index.
ISBN 978-1-55753-432-3 (alk. paper)
1. Animal experimentation. 2. Laboratory animals. 3. Human-animal
relationships. 4. Animal welfare. I. Arluke, Arnold. II. Michael, Mike.
III. Title.
HV4915.B57 2007
179'.4--dc22
                                                                  2006034333

These emotional relations, made of expectations, faith, belief, trust, which link each rat to each student, disclose the very essence of the practice: this is a practice of *domestication*. As long as this practice proposes new ways to behave, new identities, it transforms both the scientist and the rat. Both the student and the rat transform the practice that articulates them into what we may call an "anthropo-zoo-genetic practice," a practice that constructs animal and human. The rat proposes to the student, while the student proposes to the rat, a new manner of becoming together, which provides new identities: rats giving to students the chance of "being a good experimenter," students giving to their rats a chance to add new meanings to "being-with-a-human," a chance to disclose new forms of "being together."

—Despret, 2004

Contents

The Sacrifice

Introduction

Antibiotics, anaesthetics, vaccines, insulin for diabetes, open heart surgery, kidney dialysis and transplants, treatments for asthma, leukaemia and high blood pressure . . . these are just some of the major medical advances that have depended on the use of animals in medical research and testing.
— Research Defence Society, 2005

To imprison animals . . . to deliberately inflict pain, cause extreme suffering, mental distress, and ultimately a premature and often slow and protracted death all in the name of science is unacceptable. All the more so because the experiments are bad science in the first place: they do not work and have the potential to harm human health. Ending vivisection will benefit people as well as animals.
— Animal Aid, 2005[1]

These are two very different views on the use of animals in science—two poles of a fraught controversy. On one side are those who believe that medical progress depends on using animals, and on the other, those who believe that medical research using animals is not only unethical, but bad science to boot. Implacably opposed and rarely finding any common ground, neither side seems willing to back down. It seems a perennially enduring battleground.

Humans do, of course, use animals in many ways. And many of those uses are contested; hunting, intensive agriculture, circuses, for example, all have their vociferous critics, opposed to perceived cruelty. Yet perhaps nowhere is the use of animals condemned more strongly than the laboratory. The use of

1

animals in scientific research has a long history—and also a long history of opposition. Today, most people in Western countries are familiar with newspaper articles expounding upon alleged appalling cruelty of a particular lab, or the suffering of particular kinds of animals. At the same time, changing social attitudes toward other animals—a growing sensibility about their sentience and potential suffering—ensure that people ask more questions; in turn, those who do use animals must seek more justification. Whatever moral stance we take on the issue, we know the story well.

Yet whatever the rights and wrongs of using animals, it is also a human dilemma. Because it is a controversy that provokes strong reactions, laboratory use of animals presents questions in the lives of most scientists and technicians (whether or not they acknowledge it openly). Few take the issue lightly. But they do generally believe that there is a greater good which can justify the use of animals; they may suggest potential medical advances, or gains in knowledge about how bodies work. It is for that reason that they endeavor to come to terms with using animals in ways that would, outside the lab, be considered unethical. For most researchers, the use of animals in potentially painful research can be justified with two caveats: it is justified, they argue, *if* the knowledge gain is great enough (trivial knowledge is not justification enough), and *if* the animals are well cared-for and care is taken to ensure anesthesia where appropriate.

Not surprisingly, animal rights activists are by contrast deeply hostile to animal experimentation, whatever its possible outcomes. Using animals is simply wrong for opponents because they are convinced that some lab animals suffer and because they question the link to medical progress. They generally do not believe that any kind of balance between costs and benefits is possible, emphasizing that it is animals that pay the costs, while those animals can never benefit. Nor do opponents generally believe that scientists take adequate care of the animals; on the contrary, they often suggest that animals suffer unnecessarily at the hands of an uncaring scientific profession.

Believing in a greater good does not necessarily relieve those who use animals from criticism, causing them to be reluctant to express their own doubts in public. Researchers are only too well aware of the growth in public concern about animals and in public perceptions of science in general. Since the publication in 1975 of Peter Singer's *Animal Liberation* (and later, of Tom Regan's *The Case for Animal Rights*, 1984), there has been considerable growth in what we might call the animal movement. This is by no means a unified or coherent movement, and the activities in which its constituent groups engage vary from lobbying and protest marches to destroying laboratories, attacking people, and removing animals. Needless to say, this growth has been bound up with a proliferation of books and articles deliberating on the philosophical pros and cons of animal rights. There have also been several studies of animal rights activists and organizations (Sperling, 1988; Jasper and Nelkin, 1992; Herzog, 1993). There may be many reasons for resurgence:[2] but whatever contributes, it is clear that there is now much more concern about animals, and consequently more suspicion of science laboratories.

One framework particularly relevant to the controversy over research animals is the "anti-instrumentalism" identified in many new social movements by Jasper and Nelkin (1992). By this they mean an opposition to what is perceived as a persistent—indeed, growing—tendency to see nature/the environment only as means to an end, particularly by science. Environmental protesters are an example: they typically value the environment for itself, over the construction of, say, yet more roads or forest logging, which they see as embodying destructive, instrumental values. Indeed, the animal rights movement shares with much of the wider environmental movement the flavor of a moral crusade.

With regard to animal protests, that anti-instrumentalism is linked to an ever more apparent public uneasiness about science and technology, and concomitant loss of trust in scientific expertise (e.g., Lyotard, 1984; Giddens, 1991; Beck, 1992; Wynne, 1996). In response, governments in the United States and Europe actively try to promote "public understanding of science" in the belief that there is insufficient awareness of the benefits of science. This is a theme identified by many scientists we have interviewed; for them, part of the problem regarding the public controversy about animal experimentation is that "the public" does not fully appreciate what scientists try to do, because these people do not understand the science.

Yet it is far from clear that the public *is* unaware. On the contrary, people are often well aware of the benefits while being uneasy about the power of science and technology; greater public understanding of science thus does not necessarily lead to greater acceptance (e.g., Wynne, 1991, 1992; Michael, 1992; Irwin, 1995). This public ambivalence becomes particularly sharp in relation to newer biomedical uses of animals. For example, developments in biotechnology concern many laypeople, not least because of the risk of altering "nature" (Turney, 1998). Altered animals (genetically engineered agricultural animals, for example, or the production of animals with human genes to provide some product of clinical value) have a particularly salient role here. Not only do these animals represent all the issues raised by laboratory and agricultural practice in general, they offer new challenges as they threaten the boundary of what is human.

Our concern in this book is to look at the controversy around animal experimentation as a dilemma that researchers must find ways of addressing. We are not concerned here with issues of ethics, with the rights and wrongs, about which a great deal has been written. Rather, we look at it as a particular controversy which presents practitioners with dilemmas that must be confronted. We approach this question from two standpoints: the first is from our interest in the complexities of relationships between humans and non-human animals. Whatever the ethics of animal experiments, they do involve a particular form of relationship, which in turn shapes the experiences of both human and animal. Secondly, our approach stems from an interest in the public understanding of science. Whatever else it is, the controversy about animal use is fundamentally also about how the public understands what goes on in science—and also about how scientists perceive the public. It is, as we shall see, a battle fought on the terrain of public opinion.

In the context of rapidly changing sociopolitical mores about our relationship to animals, people who work with them in laboratories must find ways of coming to terms with their actions. They must do so not only in terms of their own feelings about animals and animal suffering, but also in terms of their awareness of wider public antipathy. We explore some of these issues in this book, drawing on a number of studies we (and others) have carried out on researchers and on other aspects of this controversy in the United States and in Great Britain.[3]

Despite the diverse contexts and rather different cultural and legislative frameworks in these two locations, the above studies have shown many similarities. They concur, for example, in showing how researchers and animal caretakers seldom find it easy to use animals (and often express unease), but simultaneously feel themselves to be under siege from what they see as an ill-educated and gullible public that fails to see the benefits science might bring and is overly influenced by anti-vivisectionist claims.

Noting the uneasiness of laboratory personnel should not be taken, however, as implying that we, the authors, are trying to exonerate science or scientists from challenges posed by animal rights arguments. It is undoubtedly true that many experimental procedures seem downright cruel to observers and can cause suffering. Methodologically, we are not taking a particular position on the controversy here (although privately each of us may do so), and by describing ways in which researchers talk about their ambivalence we are not intending to take any stance. Whatever our private feelings about the ethics of animal research, we are interested in the controversy and arguments: they shed light not only on debates about the significance of science, but also on our changing relationships to other kinds of animals. Our aim in this book, then, is to examine the changing contexts of scientific use of animals, and how researchers deal with ethical and emotional dilemmas in various contexts. But what we particularly want to emphasize here is that relationships with animals in the laboratory, as well as the need to confront the moral dilemmas, help to shape the identities of people who work in labs—as well as the identities of the animals who reside there. Like other relationships, the identities of humans and animals work hand-in-hand.

We also draw on what may be termed the sociology of scientific knowledge. An important point of departure for this field of inquiry is to ask why and how scientific knowledge gets to be so influential that we are inclined only to believe something if it has been "scientifically proven." How does scientific knowledge acquire so much persuasive power—so powerful, indeed, that its actions toward animals can be justified and supported within the wider society? And what happens in laboratories—as opposed to anywhere else—that creates this special knowledge? Are laboratories, in other words, special? And what is the place of the laboratory animal in the creation of this knowledge?

Most scientists view what they do as "uncovering the truth of nature," revealing the facts: that is what experiments and systematic observation are designed to do. By contrast, sociologists of science focus more on the way that facts are *created* by scientists—not only through their practical actions, but also through their ways of communicating those actions.[4] In order to investigate that process, sociologists of science "follow scientists around" to find out how the facts of science emerge from the complex daily round of activities (Latour, 1987). That is, those very activities that seem completely obvious to scientists are questioned by the sociologists. Answering the question "why did you do that?" is often one of the most difficult tasks.

What emerges from this approach is that scientists work at many levels—not only the experimental—to establish "facts." Certainly, these include the practical and intellectual skills entailed in the doing of science at the laboratory bench, but also a variety of rhetorical skills that scientists use to persuade others, both within the scientific community and outside it. A fact is only a fact after consensus; discovering something is no use in establishing facts unless you can persuade others. Scientists bring a range of resources to bear in these negotiations—from rhetoric and representations to equipment, organizations, and meetings.

So in the context of the public controversy surrounding animal experimentation, we wanted to "follow scientists around" to find out how they construe lab animals, how they understand or construct the controversy and its various participants, and how they articulate their roles and the roles of others (humans and non-humans alike). There are several ways of investigating what scientists do and say, and our own work draws on all of these; Arluke's studies in labs in the United States have used more ethnographic approaches—quite literally observing what scientists do, and sometimes participating in their daily routines, as well as following their written and spoken pronouncements. Birke and Michael, working in the United Kingdom, have focused more on following the discourse of scientists, in interviews and written papers and contributions to other texts (e.g., media, policy documents, publicity materials). What is very clear from all these studies is that lab personnel are acutely aware of the public controversy. Protests, like animal experimentation itself, have a long history and have had considerable impact on what researchers believe and how they act, as we shall see.

Historical Context

Living animals have been used in scientific research for thousands of years. The development of medical knowledge in Ancient Greece entailed dissecting many human corpses as well as cutting into living animals to investigate how their bodies worked (Maehle and Trohler, 1987). Much of our understanding of anatomy and physiology has come from these processes of exploring bodies, whether dead or alive, and whether or not we approve of such use. But to cut into a body of any sort invokes all kinds of emotions. The practice of human dissection, for instance, led to riots in the early eighteenth century, when people thought that their bodies might be the next victims of the doctors' scalpels (Richardson, 1987), while medical students still approach their first dissection with both trepidation and grim humor (Hafferty, 1991).

Cutting into the body of a living animal similarly evokes many emotions and anxieties. Over the last century, those emotions have been at the heart of the often bitter controversy surrounding animal experimentation. Just as using living animals has a long history, so too does the expression of discomfort on the part of researchers. Writing in the second century A.D., the physician Galen remarked that apes should be avoided in work on the nervous system, so that the researcher could avoid seeing the "unpleasant expression of the ape" (cited in Maehle and Trohler, 1987:15). Although this was probably not an objection based on ethics (given classical beliefs that only humans had rational souls), the practice of vivisection has been accompanied over the centuries with occasional quiet voices of discomfort or dissent. Against a background in which cruelty to animals and people was more accepted, it is scarcely surprising that those voices were few. Nevertheless, as Anita Guerrini's study of attitudes toward experimentation suggests, some experimenters did indeed find the practice repugnant (Guerrini, 1989; 2003).

Over the last 150 years, social attitudes toward animals and the natural world have changed markedly (Thomas, 1983). Science, too, has seen dramatic changes—not only in the obvious sense of its knowledge and applications, but also in terms of its social and economic organization as well as public attitudes toward it. Among other things, public unease toward science and its consequences has grown, especially in the last few decades. One area where this is particularly apparent regards the use of animals in scientific research.

Opponents of the use of animals in research usually refer to the practice as vivisection—literally the cutting into a living body (in contrast to dissection, cutting dead bodies).[5] The first major public protests against scientific use of animals took place toward the end of the nineteenth century; at this time, most animal use had been in physiological and anatomical research, which did indeed entail cutting into the living animal. Vivisection had, moreover, often been carried out in public—the French physiologist François Magendie provoked public outcry in 1824 as a result of public displays using live animals, and by his apparent total disregard for their suffering (Rose, 1995). Anesthesia was discovered, and introduced into experiments, only in the last few decades of that century; prior to that, cutting into animals was undoubtedly excruciatingly painful for the animal. Public opposition began, then, from a growing awareness of practices that were causing immense suffering—and often to animals such as dogs, which were increasingly being selectively bred as pets throughout the nineteenth century (Ritvo, 1987).[6]

As the century progressed, concern for animals grew, and new legislation to protect animal welfare began to appear (Radford, 2001). There was also growing concern for human rights, in the form of anti-slavery laws and struggles for women's suffrage, for example. These massive social changes form the backdrop for the emergence of public protests—for it was specifically the use of animals in scientific experiments which caught public imagination and became the target of explicit campaigns, often by people involved in other social issues (many leading anti-vivisectionists were also ardent feminists, for instance; Elston, 1987).

Partly in response to the scale of public outcry, the British government passed legislation in 1876 explicitly to control what experimenters might do—the first legislation of its kind in the world, although somewhat toned down from the demands initially made by the anti-vivisectionist protesters. In principle, the Cruelty to Animals Act of 1876 limited cruelty by requiring anesthesia and restricting the use of paralyzing agents such as curare. In practice, however, it also allowed scientists to apply for a special certificate if they wished to carry out restricted procedures—a loophole condition, as anti-vivisectionists were quick to point out. Meanwhile, Henry Bergh was trying similar tactics in New York—though with less success initially. Later attempts by the Washington Humane Association to push legislation in the District of Columbia also failed, not least because of organized opposition on the part of medical researchers (Lederer, 1987).

As the strength of anti-vivisectionist feeling grew, so too did the strength of opposition from researchers. The publication of an exposé of laboratory practice in London teaching hospitals by two women anti-vivisectionists led in 1903 to a libel trial (Gälmark, 2000); this in turn led to what have been called the "Brown Dog" riots of 1907. Medical students clashed in a London park with feminists and anti-vivisectionists over the statue of a brown dog, erected in memory of the brown terrier "done to death" in experiments at London University teaching labs (Lansbury, 1985). Not surprisingly perhaps, the scientific establishment responded with the inauguration of a new association, the Research Defence Society, in 1908. Similarly, the American Medical Association established its Council for the Defense of Medical Research at about the same time (Lederer, 1987).

Gradually, however, the case for medical benefits of using animals in research became stronger. Publications of the Research Defence Society in Britain in its early years (1908–1920), for example, inevitably discussed at length the arguments of the anti-vivisectionists (as, indeed, they do now). But they also put great stress on the value of animal-based research. The use of animals in the development of anti-typhoid vaccines, and the significance of these vaccines in the First World War, played a key role in developing this argument.[7]

Although organized opposition declined after the First World War, the tension between researchers and anti-vivisectionists continued. In Britain in 1926, for example, there was a highly publicized prosecution of a dog dealer who had supplied stolen dogs to a physiology laboratory in London. Not surprisingly, a resurgence of media and public attention followed (Tansey, 1994). Experimental journals, too, had to be aware of the possibility of anti-vivisectionist protest; journal editors were careful to avoid particularly provocative phrases, as Lederer (1992) has shown in her study of an experimental physiology journal.

By the 1960s, attitudes towards animal use in research began to change again. Widespread public opposition to the use of animals from pounds and shelters led the U.S. National Institutes of Health to publish a guide to laboratory animal welfare in 1963, while in 1966 the Animal Welfare Act (AWA) was passed. Although not dealing exclusively with laboratory animals, this act set out to regulate the transportation, care, and use of certain species of animals used in experimentation. Significantly, rats, mice, and birds were excluded from this protection.

The law was amended several times in the following years, and various attempts were made to promote stronger legislation. Several animal protection bills were presented to the U.S. Congress, while in the early 1990s, animal protection organizations began to challenge the AWA's exclusion of mice, rats, and birds. Similarly, legislation came under growing scrutiny throughout the European Union; in Britain, bills were introduced into both Houses of Parliament during the 1970s, without success. The century-old British legislation was, however, finally replaced by the Animals (Scientific Procedures) Act in 1986.

All of these changes in the law have occurred against a backdrop of cultural change, notably in how we view nature (the environment) and animals. In response to this, sociologists have (somewhat belatedly) begun to realize the significance of non-humans in our social worlds (e.g., Noske, 1989; Benton, 1993), pointing to their agency and potential subjecthood (e.g., Smart, 1993; Birke and Michael, 1997; Irvine, 2004). In turn, this interest has given rise to new academic inquiries, with new journals and organizations devoted specifically to the study of human-animal relationships.

Somewhat ironically, science itself participates in this sea change. There is a growing scientific interest in seeing animals as having sentience, particularly in the emerging field of cognitive ethology (e.g., Bekoff, 2002), with its emphasis on "animal mind." Some scientists have thus argued that there is a strong case for affording at least the great apes some protection in law (Cavalieri and Kymlicka, 1996), even for preventing their experimental use altogether (as is now the case in Britain). It is just this growing awareness about animal abilities that has helped to refuel the controversy. To those opposed to using animals in science, it is morally unethical because they are our kin. And it is perhaps that point that is key to the controversy—to many scientists we should be prepared to sacrifice animals in order to protect the health of our nearest kin, other humans; while to anti-vivisectionists, many animal species are close enough kin that they too should be protected.[8]

Legislative Context

Against this background, anyone who knowingly inflicts pain on an animal is doing something which is condemned, both in law and in the mores of mainstream culture. There are, of course, ways of doing so which remain within the law, such as hunting (which can often inflict great suffering), as well as science itself. People who lawfully engage in either activity may feel themselves to be beleaguered and misunderstood. During the period leading up to the act which banned hunting with dogs in England and Wales (by February 2005), the pro-hunting community sought to argue that hunting was important for their rural communities, and that town-dwellers who opposed it "did not understand" the rural way of life; they sought also to argue that killing foxes was doing good, as foxes were vermin. Arguing that opponents or outsiders do not understand the need to kill animals, while portraying the act of doing so in a positive light, is a common strategy, deployed also (though obviously in different terms) by pro-research organizations, as we will see in later chapters of this book.

Be that as it may, there are laws governing the use of animals in experiments in most Western countries. Not surprisingly, the effectiveness of such laws are disputed—many scientists consider that the law protects animals and their welfare, while anti-vivisectionists generally feel that it serves only to protect the interests of scientists. But the main point here is that the legislative context, in whatever form, provides an important framework in which the controversy is played out.

Amendments to the U.S. Animal Welfare Act (1985) have strengthened it considerably. According to the law, for example, researchers must minimize pain where possible, use anesthetics where necessary, and avoid unnecessary duplication. Moreover, the 1985 amendments also required that proposals to use animals in research should be reviewed by a veterinarian and by members of an Institutional Animal Care and Use Committee (IACUC). In the U.S., the law sets limits on how animals are cared for, but is less specific about the uses to which they are put, a role carried out instead primarily at the institutional level.

In Britain, by contrast, controls are effected primarily by statute, with legislation stipulating conditions of both husbandry and experimental procedures. The 1986 act imposes a tripartite system of licensing. Institutions where research is carried out must be licensed, and so must the individuals who do the research. But in addition, scientists must submit project proposals for licensing. The Act also requires that laboratories have a designated veterinarian. To most scientists, the legal framework works well (though they often complain about the bureaucracy). It regulates what can be done and limits abuses, they argue. Opponents, of course, dispute this, arguing that the law merely provides a series of loopholes permitting scientists to do what they want.

In general, controls over the use of animals in experimental research operate on the basis of a utilitarian cost-benefit analysis.[9] This is particularly explicit in countries such as Britain and Germany, where the utilitarian calculus is built into the law. Thus, a British inspector, acting on behalf of the Home Office, must come to a decision regarding the possible costs (to the animals, in terms of pain and suffering) against the potential benefits and the scientific quality of the proposed project. If the quality seems low, or the benefits (in, say, medical advances) unclear, then a project using animals may be forbidden or renegotiated.[10]

Utilitarianism alone, however, is not enough to make decisions about the merits of particular research. Rather, its use within a legal framework relies on the scientific community's having some degree of consensus about what constitutes good research or what counts as medical advance. To some extent, consensus emerges out of the process of socialization, of becoming a scientist. Among other things, this process involves learning to distance oneself from the animal as animal and learning to approach it more as a laboratory tool, as well as taking on board the accepted ethics of animal use in science. We shall see several examples of this in later chapters of this book.

Animal Experimentation and the Shaping of Identities

Coming to terms with the contingencies of animal laboratories in turn affects the identities of those involved; to work effectively in the face of the public outcry requires accommodating to the dilemmas. But there are other identities shaped by the controversy—those of animal rights protesters and the general public, but also, importantly, those of the animals themselves.

This book explores the ways in which the identities of scientists, the public, and the animals used in experiments construct and depend upon each other. We do not mean "identity" here as something fixed and inherent in the self, but something which develops through engaging in social and cultural relationships; that is, identities are forged *in* relationships. As such, a person's "identity" has multiple levels and must be reconstructed and maintained. Moreover, to create an identity as a scientist or a technician entails separating oneself out from the wider culture—especially in relation to beliefs about animals. Laboratory staff must learn to accept being at once part of mainstream culture, with its complex and multilayered beliefs and representations of animals in nature (which also shapes identities), and at the same time being part of the scientific culture, in which animals are transformed into objects.

In making and remaking these adjustments, scientists and technicians working with animals must appeal to a number of others—to other scientists, to government officials, to the wider public. But in pointing this out, we recognize that there is one constituency that remains strikingly absent from such discussions: the animals. To understand the human-animal relationship as it is played out in laboratories means that we have to understand both how scientists and other lab personnel create and see their relationships with laboratory animals, as well as how the animals, too, behave within those relationships. In this book, our focus is more on how humans understand animal identities (important though the animals' identities to themselves might be), because it is contradictory human understandings that underlie the controversy.

Those human-animal relationships in turn are themselves products of a long history, in which humans have socially constructed an assortment of ideas about what is "an animal," as well as the idea that there is a special kind, called a "lab animal." There are many ways that separation between animals in general and "lab animal" came about and is now maintained—both materially and linguistically. For example, lab animals are those that are specially bred and live in specialized animal units (which, ironically, are often separated from the laboratories by some distance). These animals thus become represented as somehow different from other animals (in the wild, say, or pets).

Whatever their nature, we would not be witnessing the whole controversy if it were not for the animals. So we begin with them, discussing in the first section of this book the "identities" of laboratory animals.[11] In part, their identities are shaped by evolutionary heritage; but their identities within the lab are shaped also by the way in which they enter the (human) social realm. We know relatively little about how they experience it: even rats and mice, so often used in lab experiments, are not often studied behaviorally for their own sake (although there is growing interest in studying their behavioral needs in relation to welfare—in relation to cage sizes, for example). What we do know something about is their identities in the human context.

Animals that are now routinely bred as specific strains for research first entered the labs as wild equivalents or as animals bred by humans for other purposes. We thus start our discussion of lab animals and identities in chapter 1 by looking at the broad cultural meanings attached to the idea of the "animal" that shape how we think about them in particular contexts. Just as most people tend to think first of mammals when asked to think about "animals," so too is the lab animal often conflated with mammalian ones, and the documented history of lab animals generally follows suit. We give a brief overview of how such animals came to enter labs and—crucially—how they became standardized, turned into what some have called "furry test tubes," part of the laboratory apparatus.

In chapter 2, we move from standardization of animals to their place in experimental standardization, considering how effectively that process guarantees the science and knowledge produced. Standardization is part of a wider attempt within science to control variables and is believed to improve the validity of results. Relatedly, scientists who advocate animal use do so on the basis that animal models of particular diseases provide good data for human medicine. But how much do these assumptions stand up to scrutiny? And since "making animals to order" is intensifying with the introduction of genetic manipulations, what are the implications of such animal production for their identities?

Chapter 3 turns to a more public aspect of standardization—images and narrative descriptions of experimental animals. How we understand animals is both a product of, and feeds back into, the wider culture. That is, scientific representations of animals (in, for instance, television natural history documentaries) influence what we think about them. Thus, looking at how we portray lab animals gives us a handle on how those animals are understood by laboratory workers and by the wider public. Laboratory animals are largely perceived within research as tools to achieve specific ends. But to become so, our human understanding of them must shift away from the wider cultural perceptions of animals wrought through our experience of them as pets or as wildlife observed through cameras, for example. Rather, they must become data, and lose individual significance (Lynch, 1985; Arluke and Sanders, 1996). They are ordered as tools from suppliers, as specially bred and compliant; they can be picked out from catalogs which extol the virtues of this particular species or breed for particular techniques—much as any other catalog describes commodities. This is the main theme of the third chapter, which focuses on how lab animals are represented in scientific media—texts and images, in journals and catalogs—and what those representations tell us about the practices and beliefs of scientists.

In the second section of the book, we turn to the human practitioners of science, the researchers and technicians, to explore how they make sense of what they do, particularly within their own communities. Here, we draw extensively on our research, particularly on interviews with various laboratory workers, notably research scientists and animal technicians.[12]

People's identities depend on many things, among them the social expectations of the wider culture regarding their role, and the professional cultures in which they participate. To become a scientist, for example, means going through a lengthy process of enculturation—acquiring not only skills and knowledge but also beliefs and ethics. Like other professional identities, being a scientist is both something one always is and something which at times one is not. People may define themselves as scientists and accept the ethics and practices of science, yet define themselves as something other—with different ethics and practices—outside the lab (going home, for example, to the family dog).

For scientists, their identities as researchers must first be learned; in many branches of the life sciences, that learning process requires accepting the way that animals are used, and the way that the knowledge base has necessarily relied upon animals. Once a person becomes a practicing researcher, identities are maintained, partly through participation in professional networks, as well as through the daily routines of laboratory work. Each of these identities, moreover, is embedded in layers of other networks—manufacturers of laboratory equipment and machinery, editors of professional journals, and breeders of laboratory animals, among many others

To work with animals means that scientists must reconstruct the meanings they give to the concept of "animal," from potential pet or part of wildlife to laboratory tool. To some extent, this is built into the socialization process; training in physiology means undergoing a process of desensitization to the emotional responses that many people have to the act of cutting into an animal body (see Birke, 1994). As we will see in chapter 4, facing up to the first experience of dissecting a fetal pig in high school, or to the first experience of "dog lab" in medical school, means coping with conflicting emotions. Students must find ways of adjusting, of getting through the process, to come out the other side as part of a professional community of scientists, doctors, or veterinarians.

It is not, of course, only research scientists who must learn to see animals as "tools" with all the psychological adjustment that implies. So too must animal technicians, at least some of the time. For them, the adjustment is made more complex in that they must learn to accept the animals' role as an object of science, yet simultaneously see the animals in a more naturalistic way. In chapter 5, we consider some similarities and differences in the perceptions of research scientists and animal technicians; both can feel misunderstood by a critical public, for example, although they clearly take different roles in relation to experimental procedures. On the whole, they have rather different relationships with the animals; technicians typically say that they are the ones who best understand the animals. They are also the ones most likely to single out "special" animals to be kept back as pets and not killed.

Chapter 6 then considers some of the social relationships between scientists and technicians, and the contexts in which they occur. There is clearly a division of labor, such that most technicians have a support role, rather than participating directly in creating or determining research protocols. In that sense, they are outsiders to the science, not usually being able to influence research directions; but they are also insiders, for without them, the research cannot happen. Lab workers often distinguish, too, between insiders, who "treat animals properly," and others, who allegedly do not. People make sense of these social relationships, in turn, within a wider legal and institutional framework. All of these converge to produce specific identities.

Yet whatever their identities within the lab, scientists and technicians alike must also address larger constituencies—other scientists, the wider public, governments, even the animal rights community. How they address these wider audiences is the concern of the third section. To do so, they must not only persuade skeptical outsiders, they must also deal with their own sense of themselves as somehow stigmatized by outsiders because of what they do—feeling "behind the barricades." This is the theme of chapter 7, which explores how lab workers experience and deal with that perceived stigma. Among other things, it requires them to find ways of talking about what they do, to draw at least some people into the circle. People who might be so persuaded include other scientists, who dispute data or question whether another lab used animals according to ethical guidelines, as well as what might be termed "the rational public."

In chapter 8, we continue to look at the other audiences that scientists must address. Faced with an increasingly active animal rights movement, the scientific community has begun to mobilize, to produce counter-arguments and devise counteracting strategies. These arguments are aimed partly at what might be called "expert" communities of other scientists; they are also aimed at those with particular interests in the debate (governments, for example), as well as those members of the wider public whom they consider persuadable. In addressing these potential allies, spokespeople for science must find ways of portraying anti-vivisectionists. Typically, they are demonized, portrayed as lacking in humanity (not caring about sick children, for instance, because they put animals first). Once again, individual identities are important, as people take on different identities in different contexts and, significantly, allocate others to particular identities.

Finally, in chapter 9, we turn to the general public as a whole. Scientists tend to assume that more understanding of science equals more support. To this end, many of the campaigns led by pro-research organizations provide the public with more information, which, it is supposed, will lead to more endorsement. Communication with the public, however, contains within it assumptions of what the public is and how laypeople understand science—assumptions which are open to question. As we will argue, when scientists differentiate animal experimentation from activities that treat animals less well, the public audience is not simply being informed; it is being "corrected" both cognitively and morally.

The identities of the various human and non-human actors in this often vociferous drama are multiple and changing, including scientists, technicians, politicians, animal rights organizations, the general public, and laboratory animals. To argue the case for either side of the controversy means attempting to marshal at least some of these actors and contexts.[13] Clearly, we cannot trace in depth all these networks. In this book, our primary focus is on some of the main actors—particularly scientists and technicians, the public, and laboratory animals—to see how each element constructs identities, its own or those of the other contenders. In turn, we can examine how the ethical dilemmas at the heart of the public controversy are negotiated through those identities. Among the questions we explore are these: How do the practices of science reflect the ongoing dilemmas of animal experimentation and its public controversy? What are the roles of institutions in which the research takes place? How are the identities of the animals themselves shaped by all these, and how do they in turn shape human identities? And how does the animal experimentation controversy relate to how the wider public perceives or understands science?

PART I

INTRODUCING LAB ANIMALS

One focus of this book is how people understand their use of animals in scientific work. Most scientists see themselves as addressing a specific problem rather than studying the animals used. In that sense, the animals become tools, a means to an end. Yet, whatever the scientific questions addressed, that work necessarily centers on the animal in the lab and how its place there is understood. "Animals" have a multitude of meanings in our culture, including those meanings derived from science itself, all of which impinge on how scientists—or the wider public—perceive animal experiments. Scientists, like everyone else, must make sense of what they do not only in the light of ongoing public controversy, but also in the light of the many meanings given to animals.

Animals and people are significant actors in the activity we call science. To understand what happens in labs, and how the controversy about animal use is shaped, we need to follow both these sets of actors. Animals in science are a specific instance of how animals and their meanings contribute to human identities, a phenomenon observed throughout all human cultures (Shepherd, 1996). In this section we discuss how different kinds of animals first entered the lab, and how that is tied up with cultural change, public expectations, and changes in scientific practices. The history of biomedical research, in short, is interwoven with the history of laboratory animals. The point we want to emphasize here is that the histories and identities of both scientists and lab animals have, in many ways, shaped each other.

To examine the idea of the "laboratory animal," we consider how humans have constructed animals, both literally through breeding and metaphorically through meanings. We begin by considering how animals came to be in labs, especially through breeding programs of particular kinds. From there, we consider the human and institutional context, specifically the ways that scientific practices mirror what has happened to lab animals historically—notably through growing emphasis on standardization, not only of animals but also of experimental practices. Finally, we explore the ways in which lab animals are represented, either in written accounts of experiments or in visual imagery such as advertising. What do these tell us about laboratory animals, or about how we think about them? This first section, then, explores what we might call the place of animals in the research enterprise—their breeding, their actual location in laboratories, their significance for research conclusions, and their portrayal. Exploring these roles of animals in labs sets the stage for considering, in the rest of the book, how scientists and other lab workers deal with the ethical and emotional dilemma of using them.

"The laboratory animal" is many things. Symbolically, it serves many purposes in the course of research, just as it serves quite different purposes to pro- and anti-vivisectionists. It is portrayed in stories and advertisements as many different characters, just as humans have deliberately bred it to take many different forms, even within a species. As a generic, it describes research in very general terms—"so many million laboratory animals were used in experiments." Yet it is both generalized and standardized, as we shall see in these chapters.

I

Enter the Lab Animal

> Laboratory rats may see themselves as the most cruelly
> persecuted slaves in history, but as far as people are con-
> cerned the lab rat is the Hippocrates of ratdom; laboratory
> rats are to wild rats as Gandhi is to Hitler—they are a
> separate rat race of Koches, or Pasteurs, or Salks, or Ma-
> dame Curies. In truth we *have* conquered the rat in part
> by enslaving it.
>
> —Hendrickson, 1983:217

What is an animal? When is it something of which we are terrified—mon-
sters, savage predators, wild rats in sewers—and when is it something much
more benign? Some people think immediately of mammals, or perhaps birds,
as those animals most familiar to us. Others might answer the question of
what an animal is by thinking first about the diversity of living beings classi-
fied as belonging to the "animal kingdom," with forms as different as worms,
jellyfish, starfish, spiders, or quadrupeds. Most people would assume that if
we are using this biological classification, humans are included, belonging to
the primate order within the vertebrate phylum.

Or we might think of animals as *opposed* to humans—a more colloquial
sense, but still one referring to a wide range of animal forms. Yet again, we
might answer not so much by reference to specific animal kinds but through
the ways in which we humans think about nonhuman animals. Humans are
remarkably ambivalent in the way we think about these others. In modern
Western culture, we have inherited a history which has separated us to a large
extent from nature, and thus from many of the living forms inhabiting the
earth. Most of us live in towns, far removed from nonhuman animalkind—or

19

apparently so. Not surprisingly, the term "animal" has acquired many contradictory layers of meaning; an animal can be friend or foe, food or furniture;
some animals are cossetted and protected, others are hunted without remorse.
The extent of the ambivalence is evident in myths, legends, fairytales, as well
as modern advertising narratives.

The meanings attached to animals inevitably change as culture changes.
Sensibilities toward animals altered significantly in European culture from
the early modern period onward, helped along by the growth of pet-keeping.
Pet-keeping bridged the gap between the newly urbanized and animalkind; it
brought animals into the home and permitted pet owners to treat pets as family members, and to attribute specific traits to them (Thomas, 1983; Serpell,
1986). This in turn created a cultural space for the production of pets *with*
specific qualities, through selective breeding.

Thus, not only were the meanings given to certain kinds of animals
changing rapidly but so too were their actual forms. Following the intensifica-
tion of selective breeding for elite animals (cattle and thoroughbred racehorses,
for instance) in the eighteenth century, selective breeding of dogs to suit human demands intensified in the nineteenth (Russell, 1986; Ritvo, 1987). Not
surprisingly, selective breeding of other animals followed suit, including that
of the rats and mice who were the progenitors of today's laboratory stock.

Some kinds of animals, then, are particularly potent sources of metaphor and meaning, so much so that people have bred them to take on specific
meanings—think of a lap dog, for example, fulfilling a circumscribed role
for its owner, or dog breeds regarded as fashion accessories. Because they can
mean so many different things to us, animal images are potent and recurring
symbols in our culture: they are everywhere. We load animals with our own
cultural meanings, as well as try to make them fit our cultural expectations
(Arluke and Sanders, 1996; Baker, 2000).

Science is one source of those meanings. Natural history television en-
courages us to think in quasi-scientific terms about nature, about particular
species; we are used to the dominant narrative, which turns observed animals
into subjects of scientific curiosity. It is through science that we have come to
name animals in particular ways—as exemplars, for instance, of certain species, given Latinate titles.

Yet however much we come to know about animals in the wild, we
must often tame them for specific purposes. Scientists in the field spend days,
months, or years, tracking their chosen animals in order to get a little knowledge about them. But if scientists choose to work in the lab, then nature must
be brought to them. Either this has meant using animals which are already

domesticated, or using wild animals. Neither was an easy option: using domesticated animals such as dogs and cats means confronting their "special" status and meanings within society, while bringing wild animals into the lab is liable to be dangerous. And using animals in ways liable to cause suffering presented scientists with problems of ethics as well as public relations. Thus to create many of what we today know of as "laboratory animals," especially the oft-used rodents, people had to find ways of taming wild animals, to bring them into the lab. This transformation, from wild to tame denizen of the lab, and thence to become tools of the trade, were major changes in how science worked. As a corollary, scientists working in labs now must often make a similar switch, from perceiving animals as similar to those in the wild, to seeing them as tools of the trade.

In order to understand how scientists and technicians perceive their animals, and the controversy surrounding both, we need to set the scene. In this chapter, we begin with the animals themselves, looking at how they were used in science over a hundred years ago. We will also consider the origins of purpose-bred animals, and how they came to be such a significant part of the biological laboratory. To do so, we will draw on a range of historical and sociological studies of laboratory work, particularly toward the end of the nineteenth century. What was happening in science at the time—its practices, knowledge, and organization? As we will see, the life sciences were changing in many ways that facilitated the emergence of specialized lab animals. These changes provided an important context for the mutual engagement of lab animals and humans, with consequences for shaping both. Identities of lab animals and lab humans have been closely linked from the beginning.

Who, or What, Is "The Lab Animal"?

If the word "animal" has many layers of meaning, so too does the phrase "lab animal." To many people, it is an unhappy phrase, conjuring up pictures of animals brutally treated, subjected to unthinkable atrocities. The scientists hidden behind this understanding are often seen as evil figures, taking us into a fearful future, a science gone mad. Other people might imagine the iconic white mouse or rat, as a symbol of biomedical research, or a more generalized animal as necessary to the progress of science. In this case, the scientist might seem to be a more benign figure, bringing us closer to new medical advances. Either way, our imagining of the animal in the lab means imagining also the humans who use it.

To people outside science, moreover, the phrase "lab animal" often evokes images of mammals, especially images that call upon our pity—of, for example, cuddly rabbits or newly hatched chicks or small monkeys. What the phrase does not do is make reference to the biological diversity of the animal kingdom. Few people would think about invertebrates as "lab animals," despite their undoubted importance in some areas of work (for example, the significance of the giant squid in mid-twentieth-century studies of the electrical functions of nerve axons, or the growing use of nematodes as subjects of genetic study).

Some kinds of lab animals are so well entrenched in the research enterprise that the people who use them identify themselves with that kind of animal, of whatever species. Thus, researchers in the lab of geneticist T. H. Morgan, who worked on the genetics of the fruitfly, *Drosophila melanogaster*, became known as "fly people" (Kohler, 1994) while scientists working with inbred mice identified themselves as "mouse people" (Rader, 1998). How these particular organisms were developed, through a process of generations of laboratory inbreeding, thus created not only an animal identity but a human one as well.

Writing about the development of the fruitfly as a chosen organism for genetics, Kohler writes about experimental organisms as a special kind of technology that is "altered environmentally or physically to do things that humans value but that they might not have done in nature." Some of these, like *Drosophila* itself, have been dramatically redesigned through selective breeding to become something constructed specifically for the lab, like other equipment. Others are less changed, such as frogs; but with these, Kohler argues, "the artifice resides less in physical reconstruction than in the accretion around these creatures of bodies of knowledge about how they behave and how they can be made to do useful tricks in experimental laboratories." Once animals have entered experiments, however, they increasingly "resemble instruments, embodying layers of accumulated craft knowledge and skills, tinkered into new forms to serve the peculiar purposes of experimental life" (Kohler, 1994:6–7). Kohler does not mean here that scientists necessarily see the animals only as instruments, more that once the animal has become part of the experiment, its role in the research is *as though it were* an instrument, and carrying with it layers of meaning, just as do other instruments in the lab.

Here, we explore not only how lab animals have been created literally, through decades of selective breeding, but also how the idea of "lab animals" (as tools, as somehow helping humanity in the search for better knowledge, medicines, and so on) follows suit. The production of specialized strains of laboratory animals has been an integral part of the development of biomedical science, going hand in hand with changes in the organization of science and the careers of scientists. The fortunes of animals and humans in labs is deeply intertwined. As Shapiro (2002) has pointed out, the "lab animal" is a construction, a product of the way that scientists and other lab workers obtain and take care of particular kinds of animals, as well as of a philosophy and language of science that demands objectivity.

We concentrate here on two particular kinds of lab animals—briefly considering dogs, but focusing in detail on rodents (rats and mice). There are clearly many different kinds of animals that find their way into labs (but a surprisingly small variety), and we are aware of the irony of describing this lack of diversity by focusing specifically on mammalian species. There are two reasons for doing so: first, they are lab animals whose history has begun to be documented, and on which we can draw. Secondly, because they are mammals, and relatively easy for people to identify with, both are potent icons of debate over animal use in science, and are thus particularly relevant to the themes of this book. Moreover, it is often mammals such as dogs that provoke particularly strong reactions from medical students required to use them. Few people get too excited, by contrast, over potentially invasive or fatal experiments with nematodes or fruitflies.

Dogs, however, have long been significant symbols of protest about the plight of lab animals (Tansey, 1994), while lab rodents seem to symbolize animals as an integral *part* of the lab, and are by far the most frequently used of any kind of vertebrate animal in scientific experiments. That is not to say that other animals are unimportant: primates and rabbits have also been potent symbols of the plight of lab animals, whose images are often employed in anti-vivisectionist literature. But detailed histories of these animals in labs have yet to be written. The species we have chosen—particularly rats and mice—serve to illustrate particular themes in the way that scientific research using animals has developed.

During the wave of anti-vivisectionist protest at the end of the nineteenth century and into the twentieth, dogs, cats, and rabbits were among the mammals particularly favored for physiological experiment (Kean, 1998:97), while frogs were another commonly used test organism (Holmes, 1993; Logan, 2002). Rats and mice, on the other hand, were less widely used, and there was almost no complaint about their use (Dror, 1999)—which was no doubt an important factor in scientists' selection of them to breed for laboratory programs. Then, as now, it was the use of familiar, pet animals, like dogs or cats, that led to vociferous protest.

Indeed, dogs have long been a motif of animal suffering in labs, from paintings of scientists about to vivisect a playful dog, to the statue commemorating the terrier used in London teaching hospitals (Lansbury, 1985; Tansey,

1994). So to think about how the meanings or the identities of lab dogs are constructed is a little more difficult than thinking about rats; it is very easy to start thinking about Fido rather than the abstraction "lab dog." Yet it is precisely the abstraction that students of physiology are expected to make if they are to be able to take part in laboratory training requiring them to do procedures on dogs (see chapter 4); learning to make such abstractions is a crucial part of the training process.

Dogs (and cats) have undoubtedly played a significant role in biomedical research, especially in physiology and surgery. To those arguing that medical progress relies upon animal use, the discovery of insulin as a therapy is a significant example, based on Banting and Best's experiments with diabetic dogs. Yet dogs and cats hold a special place in the affections of people in the affluent West. To many the idea of using dogs in potentially painful experiments is abhorrent—particularly so if these animals may legitimately be obtained from pounds, as happens in some countries (including some states of the U.S.). Newspaper pictures of smoking, purpose-bred beagles leads to outcry—how much more so if the dogs in question might once have been someone's pet.

Even if the animals in question have been purpose-bred for research, most people still see them as potential pets. Many dogs and cats live lives closely entwined with ours, as companion animals, and have long been a significant part of Western culture. In that role, we are confronted with their intelligence, and often grant them a kind of personhood (in some cases, treating them in ways that defy their dog- or cathood), bringing them into our networks of social relationships. Discussing how people interact with companion dogs, Sanders (1999) notes how the human-dog relationship has significant impact on how people develop or maintain their social identity, both directly (we talk to our dogs) and indirectly, in the sense that the animal mediates many interhuman relationships.

Given this "special" status, it is unsurprising that some animals evoke particularly strong reactions if they are used in scientific experiments. In Britain, for example, the nineteenth-century legislation covering scientific use of animals (in 1876) specifically separated dogs, cats, and horses for special consideration because of the strength of public opinion, and to this day, British authorities do not permit the use of pound dogs. In the U.S., too, there remains controversy over whether pound dogs should be sent to labs: supporters argue that most of these dogs would be anesthetized anyway, while opponents point out that some of these were or are someone's much-loved pet. And even among dog breeds, there are some that are more easily identified as potential pets, making it harder for researchers to use them; for example, some researchers find it easier to use greyhounds than spaniels for this reason, while the students we consider in chapter 4 learn to categorize dogs as "not-pets" so they can dissect them.

Rodents of course do not fall into the category of special pet for most people (though some keep rats or mice as pets), and there is considerably less opprobrium attached to using them in the laboratory.[1] Some scientists, like the general public, draw a line between working with lab rodents and with animals, like dogs and cats, that they might live with at home; that is, they find it easier to carry out invasive procedures on rodents, feeling unable to distance themselves enough to use dogs (see chapter 5).

One reason for this species distinction and the relative acceptability of using rodents is the widespread cultural association between rats or mice and disease: they are, in the wild, vermin. But another reason may be that, more than other kinds of mammals, rats and mice have been specifically bred *for* the lab, to create many different strains for different purposes. Animals that have been designed as specific disease models, to answer specific research questions, seem less like other animals and more like part of the lab. It is that specific and deliberate breeding program that helps move the animal from its wild progenitor to being an integral part of the science, and from being the loathed creature of myth and story to the unsung hero of medical progress.

To understand how rats and mice made the transition from wild, suspicious animals to relatively tame, docile inhabitants of laboratories with highly controlled spaces and practices, we must first understand what was happening in science. A lab animal after all is one whose life is determined by the requirements of science, whether it has been brought in from the wild or purpose-bred. Although there have been many changes in the organization of science that have had an impact on the development of laboratory animals and how scientists use them, the demand for standardization has been especially significant. As experimental biology became established in the late nineteenth century, it set standards, particularly the requirement that procedures needed to be regulated and controlled, so that experiments could be replicable. That in turn created a demand for more and more standardized laboratory equipment—and animals. Rodents in particular fit the bill, because they are mass-produced in large numbers, to specified requirements (although specific breeds of dogs are purpose-bred—beagles, for instance—the numbers are much smaller).

Standardization is a theme running through the history of laboratory rodents, marking their transition from the quite variable state of their wild progenitors to what they are today. Standardization and replicability thus not only help shape the identities of what we know as lab animals, especially rodents, but also feed back into the beliefs and practices of scientists themselves. The standardized lab animal, moreover, becomes a separate category of animal: it is not a pet, nor is it wild. Indeed, it is when people believe that these boundaries have been breached—when, for example, they think that an erstwhile pet has been taken into a lab—that they are more likely to oppose animal experiments.

Standardization or Diversity?

What has also characterized the development of standardized procedures and animals is a narrowing of the range of species and kinds used in laboratories. By the end of the nineteenth century, experimental biology had begun to gain dominance over natural history. The shift was significant, for natural history had traditionally been concerned with a diverse range of organisms in nature, whereas the focus on experimentalism encouraged a search for organisms more amenable to laboratory life. As the older, zoological, traditions ceded ground, centers of scientific authority shifted to labs, such that "new entities were constructed . . . and controlled in labs. And they were made to measure. They brought measurement *with* them, by their very structure and place" (McOuat, 2001:640, emphasis in original). McOuat argues that it was not so much that experimental biology had different theoretical grounding, but that what mattered in this shift was place. New types were developed that *belonged* in the lab, such as lab rats: they did not necessarily have a counterpart "in nature."

Increasingly, professionalism in science became linked to experimentation. And in some areas of biomedical science, such as physiology, animals were an essential part of the practice. Part of the process of standardization included bringing living organisms under control, carefully managing their breeding and environments. As this management proceeds, diversity disappears and individual variation becomes seen as a nuisance.

Physiologists at the end of the nineteenth century tended to think that individual variation was important, that it constituted significant data. How each animal (usually dogs or cats) reacted was part of the study. By contrast, in other areas of biology, individual variation was a "noise." Bacteriologists studying immunity, for instance, using mainly rodents, sought similar responses from their animals and strove to keep variation to a minimum (Löwy, 2003). Most experimental programs today follow suit: individual variation between animals is generally reduced, both through the selective breeding programs, and through the use of generalizing statistical tests. One consequence, Löwy notes, is that the study of individual variation in immune responses was simply not studied for many decades.

Diversity between species, too, has greatly reduced over the last century. Logan (1999), writing about the early development of lab rat strains, points out how, in the late nineteenth century, laboratory researchers tended to use a wide range of organisms to study a particular problem. But as rats and mice were increasingly purpose-bred, the diversity disappeared—so much so that the white rat, she argues, became a *carrier* of generality: it came to stand as a model for generalized physiological processes. She later noted that the experimental approach required "analysis, external control, standardization and replicability" (Logan, 2002:349). One part of the quest for standardization was, she argues, the increasingly close connection between measuring machines and the animal physiology they measured.

Changes in how living systems were perceived mirrored these changes in practice—control systems, derived from engineering, became the dominant theoretical framework in physiology by the mid-twentieth century. Out of these shifting perspectives emerged a demand for greater standardization in laboratory animals. Rats and mice fit the bill—an easily reproduced animal that could be turned quickly into a "standard model," of relatively uniform behavioral and physical traits, that fit readily the requirements of machines used for measurement. Accordingly, biologists began the selective breeding of laboratory rodents in an effort to promote standardization. Standardized animals followed the production of standard machines—indeed, their development was partly inspired by the establishment of machine criteria for screw threads in industry (Logan, 1999:10).

One consequence of these changes was a shift in emphasis away from diversity—using a variety of test organisms—towards perceiving the rat or mouse as a generic organism. By the 1930s, the white rat had become *the* standard for experimental psychology, for example (Logan, 1999). Mouse breeding in genetic research, similarly, was accompanied by the active elimination of diversity to create selectable traits for study (Fujimura, 1996). In the avoidance of diversity, and the search for generality, "the" laboratory rat and mouse were created. And in the process, they moved from being animals, as understood in the wider culture, and naturalistic objects of study, to being tools of the trade, part of the apparatus of science.

Lynch (1988), in a study of neuroscientists' laboratory work with rats, pointed out the contrast in scientists' speech between what he called the "naturalistic animal"—that is, the animal of common sense, the kind with which most of us are familiar—and the "analytic animal." This is the way that the laboratory animal is turned into data. So if scientists comment that "that was a good animal," they are not necessarily referring to the behavior of the animal when it was alive, but to the way that, say, its brain tissue has turned out once it has been sliced and preserved. "The animal" here refers to the product of a series of experimental procedures.

This transition to thinking about animals as data or things in turn reinforced how scientists thought about scientific generality, suggests Logan, in her study of the history of standard animals in physiology (Logan, 2002). Moreover, "it made standardization less at variance with biologists' notion of animal. Things, after all, can be made as similar as one needs them to be. Organisms, on the other hand, are diverse. Beasts that initially were standardized simply to ensure that scientists could produce comparable findings under similarly controlled experimental conditions, that is to achieve reliability, instead became sources of generality" (ibid., p. 356). By the middle of the twentieth century, uniformity was the ideal within a strain, to make the laboratory animal more like a "chemical reagent," just another piece of laboratory equipment (Lane-Petter, 1952:30; Clause, 1993; Phillips, 1994).

The Standardized Laboratory Rodent: Creating Identities

It is rather ironic that the diversity confronting biologists has become so narrowed down in the laboratory, to just a few "model" species, which in turn have become more and more like "living reagents." It is also ironic that the species so commonly reviled for carrying disease—rats and mice—should be so widely used today in the search for the cure of disease. How did this happen? Why were mice and rats specifically chosen for particular lines of research? And what was involved in the process of creating "the laboratory rat" (or mouse), such that they become so standardized? As we will see, this was a complex coming together of the specific history of certain kinds of animals, the efforts of particular people, and significant changes in the way that biomedical research was organized.

Laboratory rats—the animals bred by the million for various kinds of experimental purposes—are derived from brown or Norway rats, *Rattus norvegicus* (even when they are white), a species which came into Europe from Asia (and from there to North America) in the eighteenth century, rapidly supplanting the preexisting black rat. Part of the identity of laboratory rats, then, comes from that evolutionary history, including processes of selection that have produced an extremely adaptable animal.

Yet the identities of rats and mice are also significantly a product of human activities. Perhaps more than any other animal that has found its way into laboratories and research breeding programs, rats and mice have a long history of human vilification and pursuit. Given the association of rats (both the Norway, and the black rat that brought the plague to Europe) with disease, they have long been hunted and killed by humans.

Wild Norway rats are quite pugnacious, and colonies of them were first selectively bred for that trait during the nineteenth century as candidates for terriers to catch and kill in the rat pits (Lockard, 1968; Golding, 1990). Yet it was this idea of deliberately breeding animals for killing (selecting them for their pugnacity) that gave rise to breeding for other traits. In Britain, Queen Victoria's ratcatcher allegedly kept back animals of particular colors to breed; from these crosses, he developed animals of specific coat colors to sell as pets. Later, rat pet-keeping and breeding ("fancies") spread, particularly to working-class communities in the East End of London and northern England.[2] Breeding for pets required selecting for quite different traits, including docility, making them more suitable for use in the lab, and so colonies of fancy rats began to provide the stock for lab strains.

There were, broadly, two stages in the development of laboratory rodents as we know them. The first was the process of bringing them from the wild into the labs, via the fanciers' breeding rooms. This entailed a transformation from wild to tame, and from animals exemplifying certain species (such as brown Norway rats) to multitudes of different types, colors, and strains. It also, of course, required a transformation from being an animal that routinely elicited reactions of disgust and horror from people to becoming an animal that would represent medical progress. The second stage was one of greater industrialization, in which lab animals increasingly also became a production process and an additional scientific tool (Logan, 2001; Shapiro, 2002). But while many of these generalizations apply also to other species, rodents became particularly standardized, and now exemplify "laboratory work."

Throughout these transformations, meanings change and new metaphors arise. The wild rat of the sewers, terror of so many myths and legends and bearer of disease, becomes part of the struggle of biomedicine to conquer disease. There is considerable irony in this transformation; as Shapiro notes, that an unruly and nocturnal animal, terror of our history, has moved from the shadows into the spotlight, to become "the primary inhabitant of this highly controlled, rule-bound, broad-daylight laboratory of science" (Shapiro, 2002).

Mouse strains, similarly, began outside the lab. A few have been known for centuries (e.g., "waltzing" mice, bred in China and Japan), but—like rats—most modern laboratory strains began as "fancies" of *Mus musculus* bred by amateur enthusiasts in the late nineteenth century. Amateurs soon found themselves asked to supply laboratories. Rodent breeder Abbie Lathrop, of Granby, Massachusetts, for example, supplied the researchers whose experiments led to the creation of, among others, the C57 and DBA lines of mice early in the twentieth century (Rader, 1998). Others experimented with crossing captive-bred albino mice and rats with wild ones. The Sprague-Dawley strain, for example, began with wild rats taken from a company dump in the 1920s (Foster, 1980).

The growth of interest in genetics from the late nineteenth century on provided a particular impetus for the breeding experiments that provided the raw material for laboratory colonies. Scientists began to join the amateur breeders in experiments to select for particular coat colors or other traits, and to study their patterns of inheritance. Clearly, animals which bred rapidly were best suited for this purpose, and both rats and mice became the subject of extensive efforts to study genetic patterns. There were other reasons why rodents were quickly taken up into laboratory breeding programs; for instance, because they are altricial (i.e., they are born immature), they facilitate studies of early development; and rats particularly were thought to have strong sex drives, important to early-twentieth-century studies of reproduction and sexual behavior (Burian, 1993; Logan, 2001). Their entry into the laboratory, moreover, occurred at a time of rapid change in science, which was becoming increasingly mechanized and professionalized.

In many ways, the identities of professional scientists and selectively bred laboratory animals were forged together at the start of the twentieth century (Kevles and Geison, 1995) as scientists brought them into laboratory research, remaining intertwined ever since. Historians of science documenting the sometimes dramatic changes in the organization of science at this time have noted how specific kinds of animals might enter the laboratory but then become transformed into something akin to apparatus. It was, notes Joan Staats in her history of the laboratory mouse, a gradual transition "from pest to pet to productive element of the scientific community" (Staats, 1965:1). Mice entered Harvard's Bussey Institute either as wild animals seeking food or from suppliers of fancies, but soon became part of the Institute's work on breeding and inheritance (Rader, 1998). New mouse strains were developed which fit better with particular requirements of experimental protocols.

Gradual transition it may have been, but it was not without human help. Rader (2004) has explored in detail the development of mouse strains in American research in the first half of the twentieth century. The success of inbred mouse strains in the history of biomedical research (especially in genetics) has meant that many have claimed credit for it, she notes, but none more so than

Clarence Cook Little. Starting at Harvard's Bussey Institute, where he studied the inheritance of coat color in mice, focusing particularly on inbreeding, he went on to help found the Jackson Laboratory, in Bar Harbor, Maine, which is still a leading producer of many strains of laboratory rodents.

Rader examined the ways in which different interests—financial, scientific, and personal—came together to create conditions in which inbred mouse strains could prosper. There are, she points out, many different human (and animal) actors in this story; but the significance of Little was that he worked to stabilize the mouse material and simultaneously to link it to research questions (in cancer and genetics). He did this literally through his work on inbreeding and coat color, done early in his career, which stabilized the mice genetically; he also stabilized the mice metaphorically, through his work to publicize (and eventually to sell) the new mouse strains as useful tools for research.

Stabilization in turn creates new forms of coherence: its merits must be believed by many people for the inbred mouse to succeed as a model organism, and in turn they help to create the infrastructure that supports it—from mouse food suppliers to global markets for the mice. But, Rader argues, biological materials such as the inbred mouse achieve that kind of power through a complex system of representations, which can be driven by one particular person—with the Jackson mice, this was Little (Rader, 1999). So the success of the inbred mouse strains was partly because of the way that research developed in genetics and, later, in studies of cancer within American biomedicine, but also because of the negotiating and political efforts of Little himself.

Alongside the selection of species, laboratory scientists must also employ particular experimental methods; accordingly, they must believe in those methods, and that the chosen species is a suitable model or exemplar for the problems they study. That was part of Little's success in establishing the mice: people came increasingly to see the animals as the right ones for a wide range of research problems. Researchers must, in addition, believe in the need to standardize equipment and in the replicability of results, as well as in mathematization as a tool with which to describe results. As the standardization of inbred mice proceeded, so too did the need to standardize their housing—both the physical spaces and the way that they were taken care of by animal house personnel (Rader, 2004).

Over time, research groups consolidate around a set of problems as well as specific species; as a result, each group creates and maintains its own "scientific capital"—output in the form of scientific papers or publishable results, but also maintaining animal colonies and creating new strains (ibid.). Just as new animal identities are forged as part of that capital, so are scientists' identities, which draw not only on the identification with a particular group but also with their chosen animal—the "fly people," for instance. As a corollary, those identities are also dependent on what is *not* studied, as Bryld points out with regard to the fate of fruitfly genetics in the Soviet Union. She notes that this insect was "transformed by the regime into an icon of human wickedness and perversion," simply because it was the preferred organism for research into genetics, which flew in the face of prevailing ideology (Bryld, 1998:53). Few people were likely to label themselves "fly people" in such a climate.

Yet identifying with particular kinds of lab animals remains important. At a 2003 meeting of people working with rats (Rat Genomics and Models), speakers felt the need to argue the usefulness of the rat against the (to them) better established mouse. "Rat people," following an announcement to sequence the mouse genome in 1999, "demanded what [the director of the National Human Genome Research Institute] intended to do about a kindred rodent, the laboratory rat" (Pennisi, 2004:45). But since then, funding for rat genomics improved and rat research was boosted. Groups and individuals continue to identify themselves by the species or animal they work with, and to organize their science around it. Crucial to that identification is the selection of a particular species as a standardized part of the laboratory.

Throughout the twentieth century, more and more strains were developed, often for specific purposes—rats bred to be diabetic, for instance, or mice with immune deficiencies. Now there are many thousands of strains, with others being developed all the time for specific purposes. The website for the Jackson Laboratory, for example, currently lists new strains undergoing development for, say, research into cardiac or brain function.[3] But these developments are within a framework in which overall reduction of genetic variation is well entrenched, in pursuit of replicable results.

The search to reduce variability goes one step further with modern techniques of genetic manipulation—specific alterations to the DNA to create, for example, mice with particular genes inactivated or carrying specific genes. Maintaining rodents, however, especially for experimental genetic modification, is hazardous, as they are prone to a host of diseases. So a significant development since the 1960s has been the development of pathogen-free strains, or strains subsequently inoculated with a known array of bacterial pathogens.[4] This takes standardization a step further, as these animals must be kept carefully sealed from the rest of the world, a process which in turn profoundly affects how buildings housing them are constructed and how personnel work around them. However much we may refer to them as "laboratory rodents," they are not living in laboratories as most people would envisage such spaces: on the contrary, the animals are segregated into specialized units within equally dedicated animal houses, sealed away from potential contamination brought in by humans.

The "identities" of lab animals, at least to humans, are thus profoundly shaped by the development and demands of experimental science. What their identities are to themselves, we cannot say, although increasingly there have been investigations into the behavioral needs of lab animals, as part of efforts to improve their welfare. In some ways, this trend works against the drive for standardization in that it puts a premium on individual variation and experience, and the factors which might make laboratory living stressful for small animals (see Manser, 1992, for example). However much is put into striving for standardization and control, people themselves can introduce variability precisely because of their interactions with animals; as Dewsbury (1992) points out in relation to laboratory rodents, even apparently trivial human actions can have consequences for the animals and their behavior.

The lives of people and animals in science have been tightly linked for a long time. Today, how animals are maintained and bred, and the spaces they live in, as well as the surrounding environment, are all controlled carefully in the practice of science. The rationale for this is that results will only be valid if the conditions in which experiments take place and in which animals live are appropriate and constant; too much variation means variable and unpredictable data. That is one important reason why standardization is so important in science. But what are the implications for both animals and science?

II

The Animal Model and Scientific Practice

A small white muzzle is pointing watchfully at him [Lipsitz] from a paper crevice: he fumbles in his pocket for a carrot chunk. . . . The white head stretches, bright-eyed, revealing sleek black shoulders. They are the hooded strain. . . . [Later, Lipsitz ponders their role in experiments, concluding that: Emotionalism in rats is (a) defecating and (b) biting psychologists.]
—Tiptree, 1976:230, 231[1]

In the James Tiptree short story "The Psychologist Who Couldn't Do Awful Things to Rats," the protagonist, Lipsitz, is a scientist who is increasingly having second thoughts, at odds with his colleagues' beliefs. In the above quote, he watches the playfulness of the rats and then thinks about their use as models for humans.

Lab animals are often portrayed as "models" for human disease states. In psychology, for example, they might be used to study emotions and how drugs might act upon them—testing potential anxiolytic drugs, for example. As models, standardized animals "stand in" for human bodies: experiments are conducted on animals in lieu of humans because the former are perceived to possess relevant physiological, anatomical, or biochemical similarities with the latter, and because it is considered unethical to conduct such experiments on fellow humans. The primary assumption here is that results emerging out of animal experiments (using standard, homogenized animals) in one way or another generalize to variable humans: indeed, this is the primary practical and moral argument made by proponents of animal use in biomedical research, who draw direct connections between animal experiments and major break-throughs in human health (see, for example, Paton, 1993).

This use of the laboratory animal as model is, however, not simply a technical matter of comparability. On the contrary, as we have seen, there has been a marked reduction in the *diversity* of species used in research: mice may indeed be, biologically speaking, appropriate "models" of some aspects of human physiology or immunology, but they may be poor models in other respects—as Tiptree implies in the extract above. Importantly, these comparisons between species entail epistemic and moral assumptions about, respectively, what is being compared with what (is it really animal and human bodies?), and what benefits are forthcoming from such experimentation. Now, whether one agrees or not that medical benefits have resulted from animal use, it is undoubtedly the case that scientific knowledge has been based extensively on data obtained from animals: what we know about physiology, say, derives historically from animal experiments.[2]

In this chapter, we explore further some of the implications of scientific reliance on standardization in the production of scientific knowledge. That is, we step back from the actual production of specific kinds of animals for laboratory use, to consider the idea of a "model," and its implications for scientific knowledge. We recognize that some animals used in research are not standardized—dogs obtained from shelters, for example, or many primates used in HIV research. However, standardization is itself a standard, a benchmark in practices of science that consistently seek to reduce variability. What will emerge in this chapter is that standardization is a major accomplishment that relies on all manner of additional and often hidden activities that manage variation and unpredictability. Not only are the animals made into standard forms through breeding, but so too is their living space, the animal house, as well as the structure of the whole laboratory—all of which is aimed at reducing experimental variability. But what will also emerge is that, despite assumptions of replicability of experiments, standardization does not always ensure it. On the contrary, there are a number of ways in which variability and unpredictability always seem to creep in—thus jeopardizing claims that the data are necessarily experimentally replicable. We then consider the laboratory animal in its pivotal persona as "model," and find increasing questions about claims that animal-based data are easily extrapolated to human medicine thus shedding doubt on the pivotal status of generalizability.

Finally, we reflect on the ways in which the promise of standardization and modeling is enhanced by recent developments in biotechnology: genetic modification of animal genomes seems to indicate that animals can now be "made to order." Yet it also implies an unraveling of our understanding of animal types, whether that be species or specific kinds of inbred laboratory strains. While these developments potentially have major implications for the symbolic role of animals in Western societies, they also serve to highlight the extent to which laboratory animals, whether genetically modified or not, are thoroughly interwoven into larger technoscientific systems—systems that heterogeneously connect technical, epistemic, cultural, social, and economic factors. People, too, are embedded in these systems, which in turn structure the experiences, and hence identities, of both people and animals engaged in the practices of science.

Standardizing Space: The Placelessness of Laboratories

The standardization of animals has gone hand-in-hand with that of laboratories. Labs and animal houses are obviously sets of built structures, enclosed spaces, in which various operations crucial to science take place. However, space is both a product of, and serves to produce social relations within it—social relations that are fundamentally oriented to the generation of scientific knowledge. Accordingly, in the animal facilities that concern us here, the structure of their spaces is deeply interwoven with heterogeneous relations between people (along the lines of divisions of labor, for example) and between human and nonhuman (both animals and technologies).

While we may speak of laboratory animals, very few actually live in the physical spaces called laboratories. Rather, laboratories using animals are now usually separate spaces from those where animals are kept or bred—activities which normally happen in dedicated animal houses. Often, animals that enter the lab do not come out again. If animals are not bred on site, but are bought in from external breeders, they may have to pass through the institution's animal house before ending up in the lab. Either way, there is a very clear separation between the two sets of spaces and activities.

At the end of the nineteenth century, institutions seldom had facilities for keeping more than a few animals indoors (Kohler, 2002). As interest grew in breeding animals specifically for lab work, however, demand also grew for places to keep them. To begin with, that was often a small room adjacent to the lab; but as apparatuses and animals became increasingly standardized, so did animal-holding facilities. One Swiss scientist, recalling work in the 1930s, remembered getting rats from the animal house, "which was a dungeon-like windowless room in the basement of the institute. . . . One would don clumsy rubber boots, heavy protection gloves, and with huge metal tongs go hunting for the large animals that lived in the closely guarded room, roaming free amid an indescribable mess of manure, food scraps and newspaper bedding" (Homberger, quoted in Hendrickson, 1983:221–2). Shipping lab animals, similarly, was not initially tightly controlled, with many animals packed in fruit crates in the 1940s (ibid., 222). All that began to change, however, after 1945, especially once pathogen-free strains were developed. Animal houses, and animal movements, thus needed to be much more tightly controlled, so that by the middle of the twentieth century, most large institutions were developing specific central facilities to serve a number of labs.

Animal houses must have a number of stock rooms to house breeding colonies or to house animals for a research program; they also require designated rooms for feedstuffs, cage-washing machinery, and record-holding. Within a stock room, many animals typically live in batteries of cages, often slotted into metal racks. If surgery is to be carried out in the animal house, it will need an operating theater and facilities for ensuring sterile conditions; there may also be an associated recovery room. If colonies of animals are maintained behind barriers (that is, all the animals are maintained as pathogen-free), then special facilities are needed to manage them. Personnel may need special clothing and hence facilities for changing; they may also need decontamination rooms and showers. All these spaces must, moreover, be kept at particular levels of temperature and humidity (following legislative requirements), which in turn require specialized plants. Animal houses, then, are highly organized and industrialized spaces, in which the movement of people and animals is controlled through the allocation of specific spaces associated with specific functions.

When scientists require animals for research, they may enter the animal house and conduct procedures there (surgical ones, particularly). A great deal of research, however, entails the scientist or animal technician either taking out the living animal in a transport cage or taking a freshly killed animal (or its tissues). In some cases, the technicians in the animal house are asked by phone or fax for specific tissues, which they can then obtain by killing and dissecting, and then sending the tissues out to the labs (in one institution in our research that was several blocks away). In others, the technicians may simply pass a cage through a door. Some animals (often larger species) may undergo experimental but non-lethal procedures in the lab and then return to the animal house for recovery; but for most, and especially for small animals such as rodents, the traffic is one-way: animals leave the animal house for the experiment and do not return, while numbers of the same species in the unit must be maintained by breeding or by buying new animals.

Centralized animal units, separated from the labs, are clearly less costly and easier to maintain. They must, however, be well separated from the labs. The primary purpose of this (which might be achieved by, say, a multiple door and sterilization systems) is to prevent the spread of disease. Increasingly, it also serves to safeguard the unit from attack: most animal units now have security locks on all doors. So in addition to controlling the spread of disease, the structure and location of animal houses also serve to control the movement of both animals and people—outsiders as well as researchers and technicians. Movement across the boundary is restricted. Movement is further controlled within the spaces of the animal house. Unlike laboratories, which tend to be rather open in design, animal houses tend to consist of a number of smaller rooms off central corridors (for ease of maintenance of animal colonies). Neither people nor animals can circulate freely in such spaces.

The spaces of labs and animal houses control animals in an obvious sense, since they are usually restricted to small cages within racks of other cages; animals are also controlled through their vicarious movement. Perhaps when animals first came into laboratories, the boundary between wild and caged was somewhat porous (in that researchers sometimes obtained animals from the wild and then bred them in-house); but as science became more industrialized, that boundary became more absolute and the majority of lab animals now are purpose-bred. Their experiences of life are thus structured and determined by the structures of science from the very moment of their conception.

Spaces available to most lab animals are generally stipulated by law in many Western countries, specifying minimum requirements of floor area and/or cage volume. But the spaces within lab animal cages are inevitably limited, and animals rarely have room to do much more than turn around.[3] They are, moreover, in an olfactory field saturated with the smells of animals in other cages (which may be stressful, and is likely to affect the physiology of species reliant on olfactory cues). Visually, they can often see little but other cagemates, as barriers between cages are usually solid, or out into the spaces of the room.

Larger animals are typically housed in cages from which they look out directly at the human viewer; small animals such as rats and mice may be housed, however, in cages that slide out of the rack, and which can be accessed from above by means of a wire grille over an opaque plastic box. Animals kept in racked cages can look out only by stretching up to the wire and peering through the gap. Significantly, two technicians told us in our research interviews that scientists in their lab preferred opaque to clear cages for rodents, because in clear cages, "you could see them all looking at you," which left the human viewers uneasy.

Surveillance is built into the design of most animal houses. Animal rooms must be easy to clean, but they must also lend themselves to an easy check on the occupants, while not necessarily permitting them to look back.[4] Cage type matters too: racks of cages are arguably easier to maintain and certainly easier to check up on. Closed circuit television might further enhance that surveillance. What lab animals always experience is close scrutiny, a watchful gaze.

Interactions with any cage mates are inevitably restricted because of the size and shape of the cage; unless extra furniture is provided as "environmental enrichment," the cage is basically a square box offering little or no place to hide. So if groups start to fight, the animal house staff must separate them. Interactions with humans are even more restricted, usually channeled through the front of the cage. While animals can be habituated to these interactions and may sometimes welcome them (associating humans with petting or with food, for example), they may also find the interactions stressful if they perceive humans as potential predators (Caine, 1992). Spaces—both throughout the animal house and within the design of animal cages—undoubtedly affect the welfare of the animals and the relationship between humans and animals within. In that sense, we might say that we come to understand lab animal identities as products of confinement within controlled spaces. Those spaces are, like the animals, highly standardized.

Producing Knowledge

There is some irony in the fact that the "nature" that scientists claim to study is scrupulously kept out of laboratories and animal houses; sterilization ensures microorganisms are kept at bay, while great efforts are made to keep out wild versions of lab animals such as rodents (Knorr-Cetina, 1983). They, and the diseases they might bring, must at all costs be kept outside of the artificial space that is the laboratory. It is that separation from nature that, argues historian of science Robert Kohler, gives modern laboratories some of their authority, because it generates a sense of placelessness: labs (and animal houses) are all the same insofar as they are designed to be scrupulously untainted by extraneous nature, and it is this sameness that putatively allows for replication of results to yield such generality. Such "placelessness gives biologists a huge advantage over those who work in nature" (Kohler, 2002:473). But "placelessness" is also grounded in the standardization of animals into tools: animals in labs are no longer seen as products of local conditions (as animals in the wild are). On the contrary, scientists make efforts to control other local conditions, to render them irrelevant; among other things, for an animal, the inside of one cage is pretty much the same as another of similar size.

It was not, however, always so, and the places we call laboratories have changed much over the past century or so. Where once the laboratory was a private space in which gentlemen-scientists revealed truths (Shapin, 1988), laboratories have now become much more standardized, even industrialized, with interchangeable components and personnel. One lab is pretty much like any other. Specific local conditions, once part of the negotiation of scientific truth among witnesses to experiments, are no longer seen as relevant.

In seeking to understand how it is that science has gained such extraordinary authority in our culture, sociologists of science have drawn attention to the importance of vision and optical technologies. It is not just that for something to be believed it must be seen by a credible witness (a requirement for scientific knowledge established early in the scientific revolution: Shapin, 1988), the seeing itself must be quantified visually, to generate something else (numbers and graphs). It is these that now have come to constitute scientific knowledge, and it is these that persuade others. Moreover, at the same time as these technologies have multiplied, so the public—sometimes a witness to demonstrations of scientific experiments in earlier centuries—is increasingly kept out of laboratories.

Observing what went on in labs, sociologist of science Bruno Latour notes how he was struck by the way in which much lab practice could be characterized by "the transformation of rats and chemicals into paper. Focusing on the literature, and the way in which anything and everything was transformed into inscriptions [such as print-out from machines], was not my bias, as I first thought, but was for what the laboratory was made" (Latour, 1990:22). Scientists, he argues, are obsessed with graphs, diagrams, printouts—whatever it is they claim they are talking about. But they seldom see the animal that supplied the data; on the contrary, such "objects" are discarded from laboratories and from subsequent reports: "bleeding and screaming rats are quickly dispatched. What is extracted from them is a tiny set of figures" (p. 39). These figures are a form of visualization; what is seen are the numbers representing cells, genes, or whatever.

As these "inscription devices," as Latour calls them, multiply, they come to affect how space is organized in the lab: they physically take up room and they shape how humans use space. Further, some of these devices also influence how animals enter the space; as we noted in chapter 1, lab animals have sometimes been bred specifically to fit the apparatuses (such as stereotaxic equipment to hold the animal's head firmly, so that parts of the brain can be visually located or electrodes implanted).

What is ultimately produced materially in the lab, via various visualizations, are documents—a research paper or an oral conference report. Thus, it is the written text that is now the "public face of science"—accessible to all who can read its often convoluted forms—rather than the public demonstrations of science given in earlier centuries. In such papers this placelessness is presupposed: neither specific lab spaces nor particular animal houses are ever mentioned in scientific reports. Concomitantly, no sense is evoked of the scientists as particular embodied actors. The physical work of science, as well as the activities of the animals, largely disappear in these texts.[5] Thus, alongside the standardization of animals has been an essential standardization of labs and lab practices, which in turn generates a kind of placelessness. The knowledge that science produces, we should assume, is universal; who did it, or where, is largely irrelevant.

This peculiar production of highly authoritative knowledge has been much debated by sociologists of science. The main point, however, is that the routinization of scientific practices and the focus on "inscription devices" fundamentally shape the spaces in which it is carried out. Animal house architecture is, in practice, organized around routinized patterns of work, into which staff and animals must fit, and which in turn affect how knowledge is produced. The more routinized the processes, the more standardized the spaces, the more predictable (or replicable) should the results be. That is the underlying assumption that underpins the silence about place in scientific writing.

However, as we shall document further below, standardization is something which must be accomplished—often in nonstandardized ways. The upshot, as commentators have noted, is that scientific data from animal experiments may not be as replicable as it first appears.

Standardization and Replicability

What is not usually present in the written-down public face of science is that the generation of scientific knowledge rests on practices that are learned through apprenticeship. People must learn, for instance, how to handle animals in ways that make the experiment run smoothly. Earlier, we noted how, in the history of laboratory animal breeding, variation has been reduced amidst the push to standardize. Variation, of course, is crucially a part of the natural world (and is the basic requirement for natural selection). Yet laboratory scientists have spent a great deal of effort expunging it from their studies, while at the same time expounding the belief that what they create is a universally applicable knowledge. If universal knowledge is thus created out of the scientific process, then the results should be replicable and generalizable. But to what extent are they? Scientists, and science students, know all too well that this is very often not the case: a teacher can carefully set up a lab to instruct students in some biological process only to have the whole thing fail dismally because of some problem with the biological material.

The production of scientific knowledge using animals, however, works on the assumption that what is produced from experiments on specific animals or strains is a *generalizable* knowledge, that what we learn from research on, say, a tumor-bearing mouse can be extrapolated to other species (notably ourselves). In this quest for generalization, variation between individuals or species is seen as a problem and must be minimized. Within an individual or group, sources of variation must be controlled through the experimental design.

It is an important part of scientists' training to learn these skills, even if they must also learn that with living organisms complete control is impossible. But although the written paper is intended to convey appropriate information about experimental controls, that information is often lacking. Indeed, what scientists and other lab workers must crucially acquire during their training is tacit knowledge—the kind of "background" knowledge of what to do in labs, how to handle animals, and so forth, that is critical to the practice of science. Standardization and generalization depend on a good deal of this unwritten background knowledge in the lab, as Geison and Laubichler (2001) point out in their discussion of the "problem" of variation in the history of experimental biology.

Trying to reduce variability also requires scientists to decide criteria of normal physiological functioning, both for themselves and the conduct of their experiments, and for the outside world, in order to define what counts as *unwanted* variability. This is often not straightforward; in Pavlov's well-known experiments with dogs in the early twentieth century, the researchers frequently found huge variations in physiological response—not only between different dogs, but also within a single dog from experiment to experiment. Pavlov's laboratory sought to represent their dogs to outsiders as "normal" and "happy," while struggling to decide what responses constituted normal or abnormal (Todes, 1997). Among other things, the researchers had to find ways of minimizing the effects of dog emotions and expectations on their experiments.

Emotional responses of animals, physiologists were beginning to recognize, could easily alter experimental outcome, introducing greater variability. So a crucial part of managing animals for research into body functions was selecting the appropriate species (ones which reacted with the least emotional responsiveness), and finding ways to reduce their emotional reactivity (by familiarizing them with lab procedures and carefully handling them: see discussion in Dror, 1999).

Pavlov and his team produced dogs with particular surgical interventions (such as gastric fistulas). Pavlov invested a great deal of time and energy in standardizing his techniques and experimental protocols in efforts to create replicability within his own laboratory, and his surgically prepared dogs were expected to produce predictable results to the student of gastric physiology. But the production of these "dog-technologies" relied on a great deal of tacit knowledge held by various people—assistants who knew how to maintain the operated dogs, coworkers who socialized the neophyte researcher into the practices of that particular laboratory, and the researchers who knew the quirks of their individual dogs and how those could influence experimental outcomes. Such tacit knowledge in turn made replication of results more difficult, especially between one lab and another (Todes, 1997).

The day-to-day practices of science rely heavily on this kind of unpublished knowledge; sometimes the only way that experimental replicability can be assured is if key personnel are transferred between laboratories, bringing with them their working knowledge (Collins, 1985). There is a significant difference between published and unpublished knowledge; published literature consistently minimizes reference to animal husbandry and potential suffering (Birke and Smith, 1994), while these may be a central part of the unpublished knowledge of laboratory workers such as animal technicians. People who work with laboratory animals usually know only too well how animals should be cared for and how they react to different kinds of experiments. By contrast to the formalities of written science, laboratory workers often make verbal reference to the animals' individual personalities, to their emotions, or to how they might be affected by experiments. By doing so, they are, in effect, destandardizing the animals, turning them (back) into individual, naturalistic animals. Moreover, as animal caretakers/technicians constantly reminded us in interviews with them it is their job to ensure the well-being of the animals, and they are the ones with the knowledge to reduce animal suffering (see chapter 5).

Tacit knowledge is, however, not itself formalized—it is simply part and parcel of how laboratory workers interact on a daily basis with the animals.[6] How animals are handled, for instance, can be highly variable and the effects poorly understood. Animals will get different amounts of handling in different laboratories—there would not be time in large labs to handle all the animals frequently. Moreover, an ability to be "good with animals" may well influence the predictability of experimental outcomes in unforeseen ways. A caretaker who is unafraid and able to handle animals easily may be able to reduce variability by handling the animals in a consistent manner, while someone else could induce variability by provoking greater emotional responses in the animals. Partly, this could depend upon the way the animals were handled—rats, for example, find being cupped in the hand somewhat aversive, whereas being scratched by a human hand while in the cage is more pleasurable (Dewsbury, 1992). Small differences between handlers could well explain otherwise inexplicable differences between experimental outcomes.

By looking at data from large numbers of animals in series of experiments it is possible to see overall patterns. In one such study, the researchers used computers to determine patterns from a data set drawn from experiments on over 8,000 mice in tests of nociception (how quickly the animal pulls its tail away from hot water, a measure used in tests of potential drugs such as analgesics); of all the variables the computer tracked, the one that had the strongest effect on the outcome was the identity of the person handling the animals—even though all personnel had been trained by the same principal investigator (Chesler et al., 2002).

And not only do personnel potentially affect the animals and their responses, so too do the conditions of husbandry, such as caging—conditions which are themselves standardized in order to reduce variability. Yet variability persists. Reviewing the growing evidence that standard laboratory cages do not necessarily yield valid data, Sherwin (2004) pointed out, for example, that cages usually used for rodents do not tend to produce good data. This was the case for several animal types used as models for human disease processes, particularly in studies of motor dysfunction. He notes that scientists working on these models of neurodegenerative disease now recognize that neurodegeneration occurs much faster in animals kept in standard cages than in those kept in enriched cages (with toys to play with, for example), so that animals in standard cages may not, in fact, be a good model for human disease. Indeed, Sherwin notes a range of studies in which housing conditions have been shown to compromise the validity of data because of effects on the animals' physiology—including effects on thermoregulation, cardiovascular response, motor and sensory function, and the metabolism of toxins—that create "unwanted" variations. Replicability, it seems, is not so easy: animals of the same strain may well react in quite different ways, perhaps because of different cages, different handlers, different labs (Wahsten, Metten and Crabbe, 2003) or because of some other, less tangible factor in their differing laboratory environments. Even exposure to ultrasound from computers might affect outcome (Sales, 1988).

So it seems that however hard scientists try to standardize and to control experimental variables, unpredictability creeps in. Physiological responses of animals in standard laboratory conditions, for example, may vary much more than scientists have generally supposed, while laboratory conditions and personnel may also be sources of variability. Furthermore, the effort to reduce variability draws on a great deal of unwritten, tacit knowledge that does not appear in the experimental protocols or published research reports. Often, it is the technicians, rather than the research scientists, whose tacit knowledge is crucial, particularly where animals are concerned. Variation, it seems, is not so easily tamed, and it is precisely those poorly understood, taken-for-granted aspects of working in laboratories and animal houses that are likely to generate individual responses and variability, however carefully controlled the conditions of housing and experiments may be.[7]

What this implies is that standardization is not something that is easily achieved; on the contrary, standardization, for all its importance, is ultimately a local accomplishment that rests on the tacit knowledge and skills of those who handle the animals—knowledge and skills that, ironically, are hugely difficult to standardize. This management of variability thus rests largely on the subtle interventions of, crucially, animal technicians, whose more or less invisible work helps to give the impression of routine, automatic, or even "natural" standardization.[8] As we will see in later chapters, technicians' identities reflect this role, as they typically identify such knowledge as critical to the successful outcome of the research and to the welfare of the animals. Moreover, research scientists' identities as producers of reliable knowledge grounded in standardization thus depends upon the unseen role played by technicians in managing variability.

Be that as it may, the primary assumption of most animal-related experimental work is that the results are ultimately *generalizable* between species. For the bulk of research, this implies extrapolation from the laboratory animal to a clinical context, usually in humans—the use of animals as "models" for human disease or physiological states. Indeed, the driving force behind most animal experiments is that the results are generalizable to people—that, after all, is the primary justification made for using animals, that the knowledge gained will be of use medically (primarily though not exclusively to humans).

Animal Models?

Laboratory animals are used extensively to stand in as "models" for human disease states: outside veterinary research, where the species biology is relevant for its own sake, potential human medical benefits are the primary rationale. The use of models is also, of course, the primary target for criticism, as opponents point to ways in which extrapolation between other species and humans fails (e.g., Kaufman, 1993; Fano, 1997). How well has the standardized animal fulfilled its role as animal model?

Apart from the problems of replicability, standardization of animals does not necessarily seem to ensure generalizability to human clinical conditions. There is mounting criticism suggesting that, in practice, human medicine does not draw on animal data as much as is often claimed.[9] If so, then the assumption (and justification of animal experiments) that results can easily be extrapolated to other species may not be valid. As we have seen, the development of standardized, selectively bred laboratory animals was characterized by a drastic reduction in the number of species routinely used in investigations. In scientific terms, this does not matter if, indeed, we make the assumption that any animal (or any mammal) will do—a rat could easily stand in for a human. It matters rather more if species differences are more significant than we believe.

Species, however, are clearly not alike, even within the same biological taxa. Yet now, in many areas of inquiry, the lab mouse or rat predominates. They are cheap to maintain, but evolutionarily not particularly close to us; the reason for their use lies in the belief that across mammals (or vertebrates) there is a considerable similarity or uniformity of biological processes, such that rats, say, can readily be used as models for human disease. That is, variability in terms of how physiology functions differently in different species is played down—a point emphasized often by those opposed to animal use in science.

Despite that assumption of similarity and generalizability, there are relatively few real-life rat diseases that accurately mimic a human equivalent; so rodent models have had to be created using various techniques, from genetic manipulation to behavioral modifications. Animal models for some disease states may indeed mimic approximately the course of human disease; as Shapiro (2002) notes, the course of infectious disease such as tuberculosis in an animal model might differ in some ways from the human equivalent, but the process and course are very similar. But such validity is less apparent for animal models of, say, behavioral problems.

Indeed, several commentators have recently voiced criticism of the prevailing assumption in biomedical science that animal models can provide us with valid information about human disease conditions. Some have argued, for example, that the claims made for using animals as causal models for disease conditions are based on weak evidence or logical inference. Thus, LaFollette and Shanks (1996)[10] suggest that as a result of this logical weakness, combined with vested interests in continuing to use animals, it is difficult to say how much human medicine has benefited from animal research. Others have looked at how often clinical research actually draws on prior animal work, finding quite poor correlations. Publication bias further weakens the link between animal experiments and subsequent clinical research, since, of course, unpublished work will not be cited (Roberts et al., 2002).[11] In one such study, the authors looked at animal studies in relation to wound healing, stroke, and heart disease or failure (Pound et al., 2004). They found that the animal trials were often poorly conducted, were done at the same time as clinical trials (thus belying the assumption that one leads to the other), or the clinical trials went ahead despite evidence of harm in animal trials. All in all, they found scant use of previous animal studies in their survey of clinical studies[12]; they concluded, "if animal experiments fail to inform medical research, or if the quality of the experiments is so poor as to render the findings inconclusive, the research will have been done unnecessarily" (ibid., p. 516). Tiptree's fictional psychologist, Lipsitz, put it more acerbically: "Reinforcement schedules, cerebral deficits, split brain, God knows only that it seems to produce a lot of dead animals," he muses (Tiptree, 1976:240).

So there are certainly criticisms of the key assumptions behind using animals in biomedical research—with both the expectation of replicability and that of generalizability. Although science operates on the assumption that experiments are always replicable, there are evidently several ways in which *animal* experiments often are not. One source of the lack of replicability ironically seems to lie in the standardized conditions in which lab animals are kept—conditions which seem not always to yield the kind of control scientists expect, as well as possibly compromising the animals' welfare.[13] Moreover, many animal experiments are justified by the belief that the results will be extrapolable to humans—again, an assumption increasingly being questioned.

Questioning the validity of animal research has been a cornerstone of the arguments used by opponents. But even if it is backed by data from meta-analyses of clinical studies of the kind noted here, it is not likely to budge the belief of many scientists that animal models yield results readily extrapolable to human clinical practice. We are not concerned here with whether or not the assumption is justified; our point is to emphasize how the production of scientific knowledge—the way that it gains such authority in our culture—is tied into presumptions about the significance of laboratory animals in science. There is, in short, a culture of laboratory animal use which presumes that that is the best, or only, way to yield valid data.

Animal research may or may not be the best method of studying physiological processes, but our existing knowledge undoubtedly draws on such research in the past. Biomedical science, LaFollette and Shanks (1996) observe, has a great deal invested in maintaining the belief and associated practices, and scientists are deeply encultured into them, as we will see in later chapters.

Not surprisingly, scientists may acknowledge that particular techniques or animals may not always be particularly good models of a specific disease state, yet go on to suggest, in grant proposals or press releases, that the information derived from the "animal models" will ultimately give rise to new insights into the causes of disease.[14] That is, moreover, the driving force behind a rapidly growing infrastructure aiming to develop, through genetic modification, animals specifically engineered as "models" for very specific human diseases.

Modeling, Standardization, and Transgenics

Creating standardized animals has entailed a definite narrowing of focus (from species diversity to a tiny range of species; from genetic diversity within those species to highly inbred lab strains), but it also paradoxically generates a proliferation within that tiny range. Over the last few decades, the number of highly specialized strains of lab rodents has multiplied many-fold, and increasingly so with the use of new genetic techniques. Genetic manipulation means that breeders can rapidly produce a new "product" in the form of an animal strain that carries specific genes, or in which specific genes have been effectively disabled. Whereas breeders previously had to produce genetic variation by crossing strains of different genetic lineage, now they can be much more selective. Tiny fragments of DNA can be inserted so that the offspring differ very slightly from their parents, perhaps only in the way that a particular protein is produced. Or specific genes may be "knocked out," or disabled, to mimic what happens in genetic diseases where specific genes are not functioning. These alterations may or may not make much difference to the animal itself, but what is produced by inserted genes might be commercially or clinically important. Thus, geneticists can produce sheep who in their milk make proteins useful in clinical practice.

Rodents, however, remain the species of choice for many research programs, and there has been a "rodent revolution" (Ahern, 1995), in which these new animal variants are created transgenically as models for a wide variety of diseases. New mouse and rat strains are being developed at a phenomenal rate as models for a great many human diseases, ranging from single-gene defects such as cystic fibrosis to complex systemic problems such as autoimmune diseases. Once scientists have produced a genetically new kind of mouse, animal breeding laboratories take over, producing the new mouse as a pathogen-free strain, and then making it available to animal facilities globally.

Not surprisingly, many geneticists are enthusiastic about these developments, and new strains of "animal models" are eulogized by breeding companies and in journal editorials. One scientist commented: "This is a tremendously powerful tool because it makes data obtained from animal studies more applicable to human diseases. Often with conventionally bred mice, the models represent the mouse equivalent of a human disease. Now we are able to manipulate these animals at a genetic level and actually reproduce the disease as it occurs in humans."[15]

Creating transgenic animals that carry particular genes for particular traits (often traits expressed in another species such as humans) both intensifies the role of the lab animal as "model" and intensifies the drive toward the artificially produced animal as laboratory artifact. In that sense, new genetic technologies carry on the tradition begun with standardization. Lab animals are now or potentially will be new models for human disease, sources of organs or cells, and bioreactors that produce proteins useful to humans. These animals are very clearly intended to be generators of data and to become crucially a part of the laboratory. Even more than their "normally" bred predecessors, they are made into commercially produced definable commodities—quite literally made to order. But it can also be argued that genetic manipulation undermines standardization, in that new kinds proliferate through this market-driven process. What is different about the new genetic use of animals is what can be called a *technoscientific bespoking* of animals—the making to order of mice, rabbits, sheep, pigs, and so on. Henceforth, they can be genetically designed with the appropriate characteristics for whatever technoscientific use they are intended for. Once this technoscientific bespeaking is accomplished, they become, as it were, "off the rack": they are a reliable product with known characteristics ready to be picked from a catalog. And just like other catalogs of commercial products, there is an apparently endless array of possible kinds.

The rhetoric used to advertise these animals in turn reflects increasing commodification; advertising for Taconic Farms' transgenic rodents, for example, informs purchasers that "the Models will not be bred" (i.e., they *must not* be interbred after purchase). Not only do purchasers have to acknowledge property rights and ensure that they continue to buy these "products" rather than trying to breed them themselves, but the animals themselves *become* "the Models."[16] Transgenic animals clearly have commercial viability and can facilitate new avenues of research. But the use of transgenic animals as improved models for human disease, too, has come into question. In their critique of assumptions underpinning the use of animals in scientific research, LaFollette and Shanks (1996) note how these new developments in transgenics are deeply rooted in genetic determinism, which typically ignores contexts, both within the animal body and in its (or our) environments. "To make reliable predictions [from using transgenic animals]," they suggest, "we would ideally need a mouse with a human metabolism writ small. Merely inserting single genes into mice will not produce a human metabolism dressed up differently" (p. 188).

They acknowledge that transgenic animals may sometimes yield results useful as predictors. But, they caution, these animals can rarely be seen as causally analogous models in the way that researchers generally assume (that is, inferring that causal links between A and B work the same in different species). These issues are hotly disputed, of course, particularly around genetic manipulation of potential foodstuffs; nevertheless, LaFollette and Shanks argue, the animals are still intact biological systems, within which the gene could work in ways not yet predicted, thus undermining their validity and usefulness as models for human disease. This is not to say that genetically modified animals are not, or could not be, useful in some research programs; rather, these authors proffer a voice of caution, warning that genetic manipulation could lead to unpredictable outcomes.

Transgenic techniques, then, can intensify processes begun with lab animal standardization—a drive toward creating animals as off-the-rack bits and pieces of laboratory equipment. But, as we have seen, such standardization does not translate unproblematically into generalizability and replicability.

Implications—Generalizing from Animal Models

In the previous sections, we have considered some of the implications for scientific research and knowledge production of the drive toward standardization, focusing on the extent to which underlying assumptions about generalizability and validity are met. We want now to turn to some wider implications in order to ask about how these themes inflect with the issue of identity, not least in relation to perceptions of animals in general.

Bruno Latour has commented on what he sees as two countervailing tendencies in modernity: toward purification—that is, separation—of the social and the natural; and hybridization or translation—which "creates mixtures between entirely new types of being, hybrids of nature and culture" (Latour, 1993:10). Modern Western culture orients towards purification, trying constantly to separate off from the natural world, even while busily creating new mixtures (as in transgenic animals or plants).

Among the many implications of these tendencies is a change in meaning of what it means to be human or animal. Part of the cultural anxiety about new genetic manipulations rests on the potential to transgress some fundamental *difference* between species (particularly humans and other species); that is why, for example, people express fear of transgenics in terms of losing some kind of species essence—fearing that a tomato with one or two genes derived from fish will end up tasting fishy. The idea of interspecies difference is deeply built into our cultural understandings of animals and animality, and underlies ethical and political frameworks that justify the different ways we treat humans and other animals.

Yet awareness of interspecies similarity is also important in our culture—we need think only of the anthropomorphizing way we integrate some animals, such as dogs, into our families and homes. It is, moreover, similarity that is the cornerstone of scientific understandings of animals; classification of the animal kingdom entails grouping according to similarities and differences, and has sought to describe processes which are basic to many different life forms (the biochemical processes by which mitochondria supply energy to cells, for instance). Research using animals operates on the assumption that there is sufficient similarity between a species used in the laboratory and the human conditions for which it is a model.

It is, of course, precisely because of such similarities that treatments work at all in human medicine as it relies on animal research: insulin does generally have much the same kind of effect in other mammals that it does in humans, and it is that knowledge that enables diabetic people to use insulin derived from other mammals. On the other hand, the similarities do not always work, as many critics of animal research have pointed out (the drug thalidomide, tragically teratogenic in humans, was tested without such effects in some lab animals).

Yet at least some of the way we perceive similarity may be a product of the very processes of standardization characterizing the history of lab animals. This focus on standardization has two implications that we want to emphasize here. First, it draws attention away from a concern with animals' welfare, in that the lab animal becomes a tool; individual variation, which might have considerable implications for animal well-being, is not considered important.[17.]

Second, it is the standardization of lab animals, and thus their epistemic suitability as experimental animals from which replicable data can be extracted, that arguably "make" the commonalities. Those species whose physiology is more different tend not to enter the lab; but also because they are likely to be nonstandardized, their differences become highlighted. While points of similarity are inevitably emphasized to granting agencies, and the stress is laid on the potential utility of the research, comparability between standardized animals and humans is highly fraught and increasingly brought into question.

So on one level, then, the animal as model and the sick human are routinely rendered comparable in scientific descriptions of physiological processes.

But both are embedded in heterogeneous networks of people and activities. These, importantly, differ: the laboratory animal is necessarily part of an experimental system that, by manipulating the body and environment of the animal, *enables* the interspecies comparisons to be drawn—that, indeed, is the whole point of the exercise. The pathological human body by contrast is, at the point of diagnosis, prognosis, and treatment, always embroiled within a (human) biomedical or clinical system. What is represented as comparison between animal and human bodies is thus, in actuality, between animals-in-experimental-systems and humans-in-clinical-systems (note we do not say animal bodies or human bodies because, to reiterate, a whole set of other factors is also crucial in mediating the comparison, not least environmental-psychological ones).

Histories of medicine tend not to show how physiological discoveries have relied on such embedded systems. Discovering how nerves work, for instance, by picturing the rapid voltage change known as the action potential relied on a long process of developing equipment capable of stabilizing and recording the dramatic but tiny electrical surges. It was not only a matter of measuring currents. New pieces of equipment were added, until a whole series of different apparatuses were connected; "the investigator could skirt total insanity," comments historian of science Robert Frank, "only because as the number and variety of devices in the sequence multiplied, some of them became standardized and perhaps even commercially available" (Frank, 1994:234).

Amongst the increasingly standardized equipment lies, of course, the animal. It is embedded in a system consisting of apparatus, technique, and biological material. In this sense, the animal embodies layers of cultural and social practices (see Birke, 1999:105–111), and it is the interlocking systems that comprise the "model" for human clinical practice rather than the animal body itself, and these systems help to create the understanding of profound cross-species similarities. Relatedly, developments in genetic technologies further lock the animal body into technological systems of apparatuses for DNA extraction and amplification, vectors for transferring DNA fragments, and so on.

There is another critical dimension to the idea of "animal model." In the historical context of animal rights discourse, the animal model is of necessity a "moral category." That is, the animal model is crucial in terms of justifying an experimental system or program of research. To justify the research, scientists have to show that it will lead to good outcomes—benefits for humans—that are greater than the "bads" of animal suffering (and other costs). As such the animal model is pivotal in scientists' identities—the "model" is always not simply a technical construct but a moral one which displays that scientists are moral actors. This assumes a particular style of thought—namely cost-benefit thinking. Morality attaches not only the valuation of the animal model, but also to the process of valuation. Thus, scientists (and regulators for that matter) also cast themselves as moral by virtue of their capacities to engage in a particular style of thought: cost-benefit calculation. As we shall explore further in part 3, this is the framework within which scientists and pro-research spokespersons judge the lay public: only those publics that are able to practice this style of thought are valued, while those that can or do not (including antivivisectionist organizations) are denigrated.

Lab Animal Identities

What, then, can we say about the identities of laboratory rodents, after their history of carefully constructed breeding programs? On the one hand they are but the purified product of a complex and heterogeneous ensemble of practices, technologies, and spaces. Accordingly, they are a purely biological entity whose similarity to humans makes them ideal for experimentation. And yet, as we have argued, this comparability rests on a host of human interventions—ranging from selective breeding programs through the organization of standardized laboratory space for the production of inscriptions to the nonstandardizable skills for caring for laboratory animals. Such interventions, then, entail a dance between ever greater standardization of the animal (its current apotheosis being transgenics), and the chronically nonstandard ways in which animals are made standard through care and individualization. Within such interventions are folded other models of the laboratory rodent: as tool and as naturalistic animal. It is that ambiguity that facilitates the unease many lab workers voice about using animals; for all that standardization and control underlie scientific experiments, variability creeps in and brings with it an animal much closer to the naturalistic. This animal is harder to categorize as a tool of the trade.

There are, however, more layers of complexity to the identity of lab animals. Laboratory rats and mice are now potent symbols of scientific endeavor, and have come to stand alongside the ubiquitous double helix as icons of the laboratory in modern Western culture. Other species, by comparison, remain more ambiguous. Dogs, for example, are sometimes obtained from animal shelters in the U.S., a practice which has been justified by users on the grounds that the animals were scheduled for euthanasia anyway (see chapter 4). But this practice means that a wide diversity of animals of different breeds and sizes are used, which cannot ensure consistency of results.

In any case, certain kinds of animals in labs will always seem slightly out of place, and it is perhaps not surprising that most of the furor about animal use seems to focus on animals closer to us either genetically (primates) or domestically (such as cats and dogs). With these kinds of animals, the meanings of lab animal and animal in other contexts clash. Inevitably, thinking about these interlocked meanings also means we must think about the controversies surrounding the use of living animals in research, a controversy which in turn shapes how we think about lab animals—just as it has shaped the ways in which we keep them, largely sequestered from public view. As we will see in the next chapter, our thinking about lab animals, and thus how they are represented in scientific writing and depictions in advertising, reflects a cultural ambivalence about animals in research.

III

Representing Animals: Unsung Heroes and Partners in Research

[A] table or graph . . . is a triumph of simplification . . . not mice, but their brain cells, not the unique features of particular cells, but what all cells have in common . . . most of the properties of the actual physical objects, of mice or of men, have been discarded, and all that remain have been normalized, ideally through quantification.

—Gross, 1990:74–75

In the last two chapters, we have explored how animals in labs came to be standardized as types, and how that came to fit experimental standardization. Here, we turn to a more public standardization—the ways that animals are represented in narrative and images. Whatever their identities, all kinds of animals serve deep symbolic roles in our culture. In myth and legend, in fairytales and science fiction, in poetry and novels, in advertising and popular culture, animal images play a hugely important part. Lab animals are no exception; their images, whether present as photographs or caricatures, or absent in these forms, convey a wealth of meanings. Here, we explore some of these, focusing on ways in which lab animals are represented in contexts aimed primarily at scientific audiences—papers, reports, and graphs produced by the people who might be expected to have first-hand experience of the realities of laboratory animals.

Yet audiences are more complex than that implies: how animals are represented in the narrow confines of scientific literatures is likely to spin off into representations of lab animals in the wider culture—with implications for public perceptions of science and animal experimentation in particular. Writing about historic changes in the visual representations of animals, Burt (2001) notes the profound changes in attitudes toward animal cruelty in the nineteenth century. Where animal cruelty had previously been a spectacle, it became increasingly hidden from view—the sequestration of slaughterhouses from public view is one example, as is the decline in (and prohibition of) public demonstration of vivisection. In that sense, animals have become invisible in some contexts, even if more visible in others.

In this chapter, we will look first at how animals are, in effect, made invisible in the way that scientific texts are written. Whereas in the late nineteenth century it was still possible for scientists to write about laboratory experiments in detail (thus one writer for the popular magazine *The Spectator* in 1875 could describe skinning a cat alive in the course of an experiment—a statement and action that would certainly not be permissible today: see Brown, 1875), writers of scientific reports became increasingly circumspect. This was partly in response to anti-vivisectionist activities but also partly a product of changing cultural attitudes affecting scientists' own beliefs about what was acceptable practice.

It was not only the visibility of animals that changed, Burt suggests, but also the "appropriate seeing of the animal." We are now more concerned with ethics, yet do not see what happens out of our immediate vision. He goes on to say that how we see animals now "becomes . . . a complex act that combines a preoccupation with the humane alongside codes that sanction animal killing or experimentation in areas outside the field of public vision. Where animals are seen, they become . . . bearers of morality in the field of vision" (2001:208).

We will see examples of this in the second half of this chapter, where we turn to images of lab animals in advertising. Here, although they are sometimes portrayed as simply tools of the trade, there is also a strong trope of animals as helpers, willing participants in the research process, even saviors. What this draws on is the moral rhetoric used in defense of animal experimentation, namely that it can lead to medical advance—lab animals as heroes of medicine.

The Lab Animal in Scientific Articles

The end product of scientific research is the published paper or conference report. Scientists first plan experiments, then carry them out, and some time later, write them up. The writing process, however, does not follow faithfully what actually happened, in part because papers are written retrospectively, and partly because the paper must answer particular critics (Latour, 1987; Gross, 1990). Inevitably, omissions and shifts of focus occur. What may be omitted are details of how animals lived or died, experiments that did not quite work, mistakes the experimenter made, data that did not fit the hypothesis.

Earlier, we noted how scientists separate common-sense understandings of "naturalistic" animals from "analytic" ones—that is, animals as they have become data (Lynch, 1988). What this distinction underlines is how scientists, in their training, must learn to separate two understandings of "animal." One kind is emotionally distanced, the laboratory artifact, produced and reproduced through standardized protocols and styles of writing. The other is the kind of animal that scientists might have at home or be familiar with in other contexts, a kind of animal with whom to have an emotional connection. It is that sense of animal, and connection, which is embedded in much laboratory tacit knowledge and appears in everyday chat around the laboratory.

Not only must scientists learn to make that separation, they must also carry through that distinction when describing animals in everyday speech, and when writing about animals for scientific publications. If experimenters are working with a species generally acknowledged to be highly sentient, such as the chimpanzee, they must lead a kind of double life. Daily chat in the laboratory can well recognize the individuality and consciousness of those animals; yet, the same lab will report the experiments in the familiar, detached, language of scientific reporting (Wieder, 1980). In speech, the animals might have names, individuality; in the papers, they usually become numbers.

Ethnographic studies of scientists and their conversations while they carry out procedures underscore the disparity between day-to-day science and written reports. In one example, a perfusion failed partly because the animal was squirming and biting the experimenter, leading to much swearing at the rat. In cutting open another animal, the experimenter severed a membrane around the animal's heart. During these experiments, researchers made several stabs with syringes or had to deal with leaks in the perfusion apparatus. Yet the research reports noted merely that animals were "sacrificed under nembutal anesthesia by intracardial perfusion utilizing a mixed aldehyde fixative media" (Lynch, 1988:74).

Experimenters must acknowledge their use of animals in research when papers are written for publication. Part of experimenters' justification for detailing their methods in written reports is that other scientists can follow the procedures; in principle, the experiment ought to be replicable and the methods sufficiently detailed. But this highly formalized way of reporting leaves little room for recounting the mishaps along the way, which are routinely omitted from reports.[1] At the same time, the paper may be read by those outside laboratory science, including anti-vivisectionists: not surprisingly, scientists are sometimes wary of noting too much detail and attracting criticism. Research papers thus must tread a fine line between giving procedural details for other scientists to follow, yet not giving away too much. Scientific journals have long been wary of publishing details of animal use, for fear of reprisals, as Susan Lederer showed in her study of the *Journal of Experimental Medicine* between the 1920s and 1940s (Lederer, 1992). Editorial policy at the time insisted on removing references to any procedures which might particularly provoke anti-vivisectionist wrath, and altering phrases and experimental numbering systems (so that the severity of an experiment or the numbers of animals used seemed to be less).

Scientific writing has become increasingly formalized, particularly since the nineteenth century, and typically employs passive, highly stylized language, often obscuring what actually happens. Significantly, the passive voice removes the experimenter from the sentence. Research reports are written so that "agency is left implicit, and if we read the passage literally we get the sense that a 'lesion' or ' sacrifice' is accomplished through an autonomous technology freed of the vicissitudes of human agency" (Lynch, 1985:151–52). In written reports, moreover, active verbs occur almost entirely in connection with intellectual processes (Bazerman, 1988). The scientist may actively generate hypotheses, while the animals are miraculously injected by an invisible hand.

The passive voice is coupled with the use of euphemisms and the frequent omission of details of how animals live in laboratories, which combine to obscure what happens to animals. There are various euphemisms, but the most obvious is the word "sacrifice" rather than "kill" in scientific reports, drawing parallels with ritual sacrifice (Arluke, 1988; Lynch, 1988; Phillips, 1994). This gives the killing a symbolic importance, as though the animal were sacrificed for some greater good. Although more papers now use the less ambiguous word "kill," "sacrifice" is still widely used, carrying with it many other meanings—hence the title of this book.

Scientific reports tend to use the word "kill" more often if the animals' deaths are explicitly stated. Perhaps surprisingly, however, deaths are not always stated. Some merely refer to the animals' being sacrificed, without any details of their deaths. But if the paper reports experiments using terminal anesthesia (in which the animal is not allowed to recover, but is given an overdose of anesthetic), no euphemisms appeared: on the contrary, the animals' deaths were simply not noted at all (Birke and Smith, 1995).[2]

Details of procedures done to animals are sometimes downplayed. Given space limitations, details of what happens to the living animal get less consideration than what happens to its tissues after removal. Then, the tissues go through various processes of preparation (depending on the experiment), which require technology—they may be centrifuged, or fixed and stained, for example. At this point, fine details of the apparatus are usually given—model, running speed, and so on. These technical details are, of course, crucial if the experiment is to be replicable. But so too are the details of how the animals were maintained—cage size, temperature, or group sizes, for instance, all of which might affect the experiment's outcome—yet these details are often omitted (Smith, Birke and Sadler, 1997).

The experience of the animals might, furthermore, be minimized by descriptions of methods of measurement and surrounding apparatus. For example, one paper referred to the preparation of rats for chronic recording involving electrical wires that were "exteriorized at the head, and the connector was bonded to the cranium."[3] By using such phrases, the writer subordinates what happens to the animal to the descriptions of what was done in order to take measurements.

Another kind of omission is that of mistakes, like severing an artery or injecting the wrong substance. To become part of data, part of the experiment, animals have to die as part of the experimental procedures; animals that merely upped and died or were killed inadvertently cannot count as data (Lynch, 1988). So animals that cannot contribute to results must be accounted for. But only rarely is the mistake explicitly noted as such.

The difference between reporting and not reporting such events partly depends on the design of the experiment, and the stage in its execution at which the mistake occurred. Thus Lynch (1985) described one instance in which a researcher had ruptured an artery while lesioning a rat's brain; because he succeeded in stemming the bleeding and doing the lesion, the animal would be included as data and, presumably, the accident not recorded. But in other instances, the writing might gloss over mistakes: "the final N was less by 1-2 in some cases due to an inability to assay every sample," claimed a paper analyzed in one of our studies (Birke and Smith, 1995), without specifying just what was meant by "an inability to assay." Again, the passive voice glosses over; there is no person who went off for coffee or otherwise failed to do the assay on time, merely an "inability to assay."

Clearly, what we have outlined here are generalizations based on analyses of several journals. Others have drawn similar conclusions from reviews of scientific writing. While it remains generally true that details are often omitted (particularly details of how animals were housed), many journals do manage to provide such detail. Space is always at a premium, so including all information is not always possible; yet some journals do manage to include such details as housing size, or stocking density. Thus, journals specializing in the scientific study *of* laboratory animals or their behavior are more likely to give details of animal housing and maintenance. Such journals focus on the animal itself, rather than on its tissues or organ function, so greater attention to husbandry details is not surprising.

Although it is not explicitly stated, the formalization of scientific writing over the past century—through which the presence of the living animal is obscured in the writing—is partly a response to criticism of the use of animals. This may be unconscious—after all, scientists are trained to write in a particular way—or authors may play down what they did to animals because of their own ambivalence or fear of reprisals. This is not to say that scientists are deliberately obscuring controversial details. On the contrary, scientists are usually well aware of the dilemmas involved. But they are trained to make a separation between animals in different contexts, to attain a dispassionate stance, and to write up the experiments according to specific patterns.

The rhetoric of science, and the peculiar and particular construction of the scientific paper, have evolved over centuries. Obscuring animals is one consequence of that developing style; it may also be a consequence of other changes in the practices and narratives of science. Science claims to study nature; yet what enters the laboratory (including the animals) is highly artifactual. The raw materials of science are "carefully selected and 'prepared' before they are subject to 'scientific' tests. . . . To the observer from the outside world, the laboratory displays itself as a site of action from which 'nature' is as much as possible excluded rather than included" (Knorr-Cetina, 1983:119). There is a certain irony here; as we noted previously, wild rodents—part of "nature" and bearers of disease—must be excluded from laboratory animal houses while experimental rodents are carefully and selectively bred to exclude disease and to make them as unlike their wild counterparts as possible (Herzog, 1988).

Animals in Laboratory Advertisements

Although the style of formal papers tends to write out the animal and its experiences, animal images appear nonetheless in a wide variety of contexts in scientific journals. In that sense, "nature" is less obviously excluded, at least as provider of beautiful images. Animals appear, for example, in editorial comments on animal-related issues in broad-based journals such as *Science* or *Nature* (articles on animals in the wild, for instance, or on the animal experimentation issue itself). In these contexts, photographs might appear. But animal images appear most often in advertisements in such journals, to sell all kinds of laboratory equipment. Here, images of wild animals "in nature" frequently appear, even accompanying advertisements for unrelated technologies, although the animal is incidental to the product advertised. Rather, the image is used to link the technology and the textual narrative. Thus, pictures of, say, a jaguar or a cheetah may be used to get across the message that this or that piece of equipment is super-fast, while advertisements for systems designed to promote protein expression might use an image of a bright blue poison dart frog to convey something of the complexity by which proteins are expressed in the frog's bright colors.[4]

Images of laboratory animals, however, occur less commonly. The most frequent images of lab animals can be found in advertisements for suppliers of equipment for such animals and for breeders of the animals themselves. Yet here images of the animals behaving normally and in their lab setting are rare, just as they are in written scientific papers. What predominates instead are abstractions (often as line drawings or cartoons), to create fantastic creatures bearing little resemblance to the real thing. Here, we will examine some of these in order to ask why such images predominate—what do they say about the scientific readership of these advertisements? And what do they communicate about animals?

Unlike scientific reports, companies producing animals for scientific research advertise openly and in doing so, help to create meanings. We saw in the last chapter how certain kinds of lab animals, especially rats and mice, have become highly standardized over time to create particular norms of what these lab animals "should" be. At times, they are referred to as though they are tools of the trade. Animal breeders in turn market this norm in their ads, trying to convince researchers that their animals are somehow superior to others. But each company is producing animals that are similar—pathogen-free lines, specific models for specific diseases, genetically equivalent strains. So the advertisers have to find other ways to attract scientists' attention.

One way of doing this is to use metaphors from everyday life and anthropomorphic images. In his study of advertisements for laboratory animals in specialist periodicals such as *Laboratory Animal Science*, Arluke (1994) found three kinds of images—animals portrayed as consumer goods, classy chemicals, or team players/helpers in research.[5] The result, he suggests, is to create an image of a fantastic animal that is simultaneously object-like and human-like, a thing both of science and of everyday life. More recently, with the development of gene mapping technologies and the publication of the full genome for both mouse and rat, a further metaphor has become common—that of mapping (Birke, 2003).

Consumer Goods

Advertisements for lab animals often imply that the animal is like any other manufactured item, meeting the specific consumption needs of scientists. One way ads do this is by making reference to breeding animals of particular size to reduce overheads, or by offering to "design" animals for specific needs. While photographs of real animals are sometimes used in these ads, the captions and text often suggest otherwise, perhaps alluding to other manufactured items, such as automobiles. An ad for a hairless strain of guinea pig, for example, was captioned "Now available in standard. And stripped down model," before explaining that "You can now opt for our standard model that comes complete with hair. Or try our new . . . stripped down, hairless model for speed and efficiency."[6]

Another strategy implying that lab animals are goods to be consumed uses drawings or cartoons to transform them into ordinary items. This might be done, for example, by using jigsaw pieces of an animal image to suggest that the breeding company is in the process of assembling a completed animal; thus, one ad ran the caption "Building a better beagle" with a drawing of a child building a wooden dog with hammer and nails. Supermarket shopping is also implied, with illustrations suggesting a shopping basket[7] containing purchasable animals.

It is hardly surprising that advertising should imply that animals are consumer goods; a glance at web pages for any of the large companies producing laboratory equipment and breeding animals shows that there is a listing for "Products and Services." This, in turn, will then list animals among other products, and then specify the different kinds of animals (bred with specific genetic defects, say, to provide a "model" for a particular disease) or the ways in which the company can provide "surgically prepared models."

In these kinds of ads, and descriptions of them as "products," the animals do indeed become "tools of the trade," inanimate objects—"models" not only in the sense of being stand-ins for human diseases, but also in the sense of being constructible, like the child's wooden beagle. These animals have, as we noted in the previous chapter, been bred for the purpose, created over many generations, and it is that deliberate production process which is invoked in this style of advertising.

The Classy Chemical

As with all product advertising, companies compete; so the animals need to be depicted not only as products, but as particularly classy or superior ones. The lab animal suggested in many ads is much more than an animal that is microbiologically clean; its purity seems to capture an essential biology of the true animal that cannot be found in nature—typically depicted as a truly "superior animal." This image of purity is reinforced partly by what is not in the ads; humans, for example, are rarely in ads for lab animals, their absence creating a sense that the animals are somehow pristine, untouched by human hands (reminiscent of the lack of agency conveyed in scientific articles). Only 9% of the ads in Arluke's (1994) study include humans, despite the fact that these animals will be used by research personnel, in many cases being handled frequently. And if people do appear, they are almost never shown conducting experiments with animals, even though this is the animals' purpose, further reinforcing the pristine presentation.

Most of the ads[8] have a photograph or drawing of the animal in blank space, without cages or other lab furniture for context. Ads often use black or white featureless backgrounds, with the animal front-lit (creating a reflective surface if this is a white rodent). In such a context, animals appear as purpose-less objects. While animal and human are rarely together in the ads, the use of anthropomorphic images also suggests purity, perhaps by using humor. One breeder claimed, for example, that its pathogen-free mice are "real health nuts." Human attributes may be accorded to animals to advertise their superiority, referring to them as "blue-blooded" or "royal." In one ad, a mouse is seated between a beaker and a flask as it composes a poem: "I'm not a common house mouse! I'm a toute a fait (entirely) research mouse . . . really upper class, I don't wear a dirty coat, nor talk to dirty people, I live in an unbreakable barrier, I'm all for quality control, because I'm caesarian-delivered, barrier sustained and truly SPF (specific pathogen free)."

Superiority might also be implied by linking lab animals to champion animals in other contexts—thoroughbred racehorses, for instance. Alternatively, purity could be emphasized by a play on more questionable backgrounds, with headlines asking, "Do you really know where your lab animals are born and bred?" thus implying that only by buying from this particular supplier can you be assured complete genetic purity. "If you don't know their background," asks an ad for purpose-bred beagles, "can you trust your results?" The ad goes on to say that problems arise from using random-bred dogs (such as those that might be obtained from a pound)—they would have "Unknown histories. Genetic inconsistencies. In a word, variables."[9]

Variables, however, can be reduced, the ads imply, by purchasing highly uniform, standardized animals. Such uniformity might be conveyed by featuring multiple images showing similarity, or even of the same animal. Thus, one breeder's ad stressed the consistency of its guinea pigs by placing the caption "Time after time after time . . . " above a dozen identical photographs of the same guinea pig,[10] thereby implying not only the similarity of the animals but the predictability of experimental results. Growing demands for genetic standardization similarly prompted one animal supplier to advertise a new Genetic Standard Rat, with a photo of a white rat (lacking background) and the caption, "If you've seen one, you've seen them all."[11]

The Map

Genome mapping has been at the center of international research efforts recently, costing billions of dollars. In March 2004, the rat genome was published, to great acclaim, following the publication of mouse and human genomes. With this flurry of mapping activity, it is not surprising that mapping metaphors surface in advertising scientific products. It is not only making maps of the genome that is conveyed in these ads, but also a kind of globalization. As Haraway (1997) has pointed out, these ubiquitous mapping images carry with them connotations of global conquest. This is particularly clear in one ad which depicts a young white woman, arms raised, superimposed on a map of Africa.

Thus in the advertisement for the Genetic Standard Rat, the white rat image appears below a map of the world, telling us that the new GS rat is an "ideal solution for the global research community," providing "reliability on a worldwide basis." A frequent trope is to superimpose an image of a lab rodent onto some image that implies DNA analysis—a double helix, or gel electrophoresis (the typical "bars" of DNA analysis). In one such ad, the white rat's body is further morphing into diagrammatic isoclines—implying that the rat's whole body is being mapped.[12]

The rhetoric by which the new mapping was announced included a number of allusions to global exploration early in the scientific revolution. According to press coverage of the genome publication, the mapping would "make the laboratory rat an even better tool for fighting human disease . . . armed with this sequencing data, a new generation of researchers will be able to greatly improve the utility of rat models and thereby improve human health." The report goes on, "This species was best known in the past for infecting ships. . . . Having the rat genome along with the mouse and the human allows scientists to triangulate, just as mariners triangulate to navigate using the stars and the sun"

(Fox, 2004). Elsewhere in the media coverage, there is reference to "Renaissance Rat," referring to the revival of interest in rats with the mapping project. With all these hints at earlier global exploration, the mapped rat becomes not only a global research tool, but also comes to symbolize the victory of Western scientific progress and conquest. The rodent, furthermore, *becomes* the map.

The Animal Helper

Perhaps the most consistent trope in lab animal advertising is that of the animal as team player or important helper in the research. Indeed, "The rat's contributions go to the heart of biomedical research—quite literally, as it is the most significant model for human cardiovascular disease. It's a similar story for diabetes and arthritis, and many behavioural disorders" (Abbott, 2004). Some ads capitalize on the docility of a particular strain by emphasizing that the animals are "on the side" of the researchers. They might be depicted next to a piece of technological apparatus, as though directly involved in using it, or looking at a graph. Some ads use cartoon images, stressing the animal's willingness to help the research (significantly depicted as a subservient female, or a servant in the advertisements Arluke analyzed).

Ads also use images implying that animals are actually part of the research team. One company, for instance, used a logo featuring a drawing of a rat wielding a syringe as large as its body. Other ads depict rats or mice dressed as corporate executives or wearing physician's whites with a stethoscope draped around the neck. Accompanying text continues the theme of research team player, by suggesting that animals "help arthritis sufferers" or are "stalking cancer." In this guise, they might be portrayed as winners of a race or as heroes; Arluke cites one example of an ad showing a framed photo of a guinea pig, with the caption, "The Unsung Hero of Bronchial Research." Another ad, for a laboratory animal breeding facility, depicts a range of animal species, with the text: "We are more than just animals—we're your partner in research."[13]

Lab animals as crucial helpers of the research enterprise is a significant theme, too, in media coverage of scientific developments. Newspaper accounts of new developments related to xenotransplantation frequently implied that animals were "helping," as though the transgenic animals used in such research were helping either the researchers or the patients themselves (Birke, Brown and Michael, 1998). These animals may "help to save lives" by producing specific proteins for use in human medicine. Similarly, an Internet publication from the Jackson Laboratory (winter, 2002) asked, "How can mice become warriors in the fight against terrorism? By helping researchers understand the genetics behind why some animals resist anthrax infection."

These animals, then, become our saviors—they help research, they enlist in global battles. The picture of laboratory animals as rescuers is a powerful one and figures in many images. One review (Paigen, 1995) of "mouse models" began with a heading, "A miracle enough: the power of mice," before going on to outline ways in which mice were ideal animals for genome research, potential saviors in the fight against human disease. It is the creature, the text claims, to whom we turn experimentally because it is "so important in reaching an understanding of ourselves."

The Fantastic Animal

Images of lab animals in advertisements allow distancing, particularly when they are portrayed as "classy chemicals," consumer goods, or perfectly obedient workers. As such, they become the quintessential scientific commodity or object that can be transformed into abstract, interchangeable units or data. But at the same time, the animals in the images can be identified with, especially when the animal is portrayed naturalistically or as a helper. Rather than being alienated from the products of their labor and the commoditization of animals, such identification may connect researchers to lab animals, at least at a symbolic level.

Furthermore, casting the lab animal as some kind of savior speaks not only of researchers' ambivalence about using animals but also of a deeper cultural heritage of salvation. In many ways, purpose-bred laboratory animals have been created to bear our diseases—from animals selectively bred to have little or no functional immune system to those who have been genetically engineered with human genes. In the case of rats and mice, they have been transformed from bearers of disease to benign assistants in the medical fight against disease. In that sense, suggests Haraway (1997), they become symbols of Christian salvation stories. The history of Western science itself draws heavily on an iconography of salvation (Midgley, 1992), so it is perhaps not surprising that laboratory animals—the sacrificed—become such symbols, bearing our suffering for us.

The caricatured animal that appears in many advertisements is highly anthropomorphic—the image is intended to convey human attributes and abilities. As such, it is clearly removed from the standardized animals housed in identical cages in the lab. Pictorially removed from its context, it is also sometimes domesticated, made to look more like the pet rabbit or cat (so blurring the boundaries of the concept of lab animal). At the same time, those animals used in science that are not standardized or purposely bred, such as pound dogs, are obviously not going to appear in the advertising images. Even ads for products other than animals themselves tend not to show such variable creatures. So the fantastic animal that appears in advertising is almost always based on highly bred standardized animals, perceived as part of the apparatus of science, but simultaneously converted by the narrative and imagery into a quasi-human.

Yet why would advertisers use anthropomorphic images or fantasy animals to gain readers' attention? One reason, of course, is that the same kind of images that appeal to scientists also appeal to everyone else—anthropomorphic images of animals are widely used within our cultures. But use of the fantastic animal speaks to deeper concerns shared by researchers who experiment on and kill animals. Writing about sacrifice in its wider context, Bakan (1968) suggests that those who sacrifice life are involved simultaneously in an act of righteousness and wrongdoing that requires them to distance and identify with the sacrificed. On one hand, Bakan contends that the killing must be seen as an external necessity for a higher being or institution; the killing is then the result of bureaucratic obedience and order. That which is killed is symbolized as an inanimate object. But on the other hand, Bakan maintains that in order for sacrifice to work, the sacrificed must also become part of the self of the sacrificer. By humanizing the victim, it becomes a surrogate self that can be identified with and seen as similar to the killer. That which is killed, then, is also symbolized as the living.

In the context of the laboratory, sacrificers—the users of animals—seem to both distance from and identify with the animals killed. As reflections of the contemporary ideology of science and laboratory culture, these animals have been found to symbolize impersonal scientific objects that can be transformed into data (Lynch, 1988; Arluke, 1990). But experimenters also often view animals as sentient beings, sometimes treating them and speaking about them as though they were pets (Arluke, 1988). Indeed, personalized relationships with lab animals are extremely important for some researchers to maintain, even if they cause uneasiness (Arluke, 1990). The figure of the fantasy animal in laboratory product advertising reflects that uneasy ambivalence between seeing animals as both part of nature and as tools of the trade.

Emerging Identities and Animal Representations

The ways in which lab animals are represented in written reports and advertising images tells us something about how animals are perceived within the scientific community: at the very least, they are viewed ambivalently. But these representations also influence how people perceive lab animals and their use, both as outsiders (the lay public, for example) and as insiders to science. In that sense, these representations contribute to the identities of people working within the labs and how they understand the animals they use.

In becoming part of the research community, scientists must learn to take many things for granted. As part of the wider culture, they have inherited a highly complex but also highly ambivalent set of attitudes and beliefs about animals that tell them that they should respect animals and at the same time be able to use them. To fit into the scientific community, scientists must learn to confront the special case of laboratory animals. But this case is somewhat different from the general cultural ambivalence, for it involves carrying out procedures that they would not consider doing in other contexts, and it draws public hostility.

However much scientists might have to deal intellectually with the need to use animals—by recourse to a cost-benefit analysis, for example—they must still deal with it emotionally. But emotional ambivalence is very clear from the ways that animals are represented in the literatures written and read by scientists. As we have seen, details of the animals are often written out of scientific reports, while advertising seems to play out the tension between seeing animals as sentient and seeing them as just another part of the laboratory.

What, then, can we conclude about the shaping of identities from our discussion of how animals are represented? There is, as we have seen, a consistent pattern by which animals disappear in written papers. Even in some of the journals which focus on whole animals (such as those concerned with studies of behavior or ecology), photographs or illustrations of whole animals are rare, especially in laboratory-based studies. What illustrates most scientific reports, instead, are graphs and tables: these are the outcome of science, helping to create its authority (Latour, 1987). As the opening quote illustrates, the mice themselves have been discarded, taken over by the graphs. The texts, too, exclude animals and their experiences through various means of omissions and circumlocutions, not least being animal agency.

If this glossing over angers anti-vivisectionists, it would not be surprising. But it also reflects the ambivalence of people who work in science and who use animals, and the broader cultural disappearance of animals from our view in some contexts (Burt, 2001). This disappearance of animals is apparent in the style of scientific reports. Firstly, the invisibility of animals in written reports renders animals as "tools"—objects. Advertising images, while not always so obviously making the animal invisible, also convey a sense of the animal as object through associated written or pictorial narrative. Thus, they create a more subtle kind of "invisibilization."

Treated as objects, moreover, animals become the site of contention when arguments among scientists arise over whether they were properly used. In this respect, standardization is at stake—so when they dispute the technical practices of other researchers, scientists are in effect "destandardizing" the animals. As we note in chapter 5, when talking about other researchers (from other countries, perhaps, or other research groups), lab workers tend to portray others outside their own sphere as being less competent at handling animals "properly" (that is, in a humane way); one implication implied by this allegation is that the experiments will not work as well if the animals are not properly handled (that is, the science is less good because extra variability is introduced). In this sense, animals thus serve as the opportunity for scientists to display not only the facts but also their own competence (this is not to denigrate science or scientists—this controversial dimension of science is simply part and parcel of its character, as sociologists of scientific knowledge have long documented). In that sense, scientists' identities reflect the tension between standardization and its unraveling.

Secondly, the invisibilization of animals in writing and advertising is a way of screening out the relations between scientists and animals—or rather it is a reflection of the conventions that scientists have learned in dealing with laboratory animals, which are no longer regarded as naturalistic animals, but rather as a source of data. Whatever interactions lab workers (both scientists and technicians) have had with animals in the past or when conducting a particular experiment, these disappear in the way that "animals" have become something else, particularly in the written or spoken research reports that result. Here, then, we see scientists' identities played out in terms of their apprehension of animals as part of experimental protocols—an apprehension that is always liable to erosion or breakdown because scientists are, of course, part of a wider culture which holds other sorts of conventions about the nature of animals. Animal technicians occupy a kind of middle ground; for them, animals are always partly naturalistic and partly sources of data for the science going on in the labs.

There is a third implication that the form of scientific representations of animals holds for identity. The disappearance of animals from such texts sets up a boundary between "science" and "public." Partly, this is in the sense of obscuring the role of animals in order to avoid wider public controversy and criticism (in which the public might be identified as irrational). But it is also in the sense that the ethical standing of laboratory animals has *already* been technically calculated. That is, experiments on animals have usually been ethically approved (at least in most countries in Western Europe and North America) prior to being carried out: so it is assumed that putting additional details about animals and ethics into research papers would open up an ethical problem that has already been solved institutionally (and this is sometimes referred to, very briefly, in a footnote referring the reader to national ethical regulations which have been adhered to). Thus, scientific identities are embedded within institutional processes such as ethical committees to which the public are not party (or party to via legislative and regulatory systems). The identities between lab workers and public, in this instance, at once converge (are all part of a system of ethical regulation) and diverge (scientists must explicitly function within it, the public has its proxies oversee the system). Furthermore, they embed the identities of the animals and their welfare needs.

Constantly negotiating boundaries between ideas of animals as naturalistic—shared with the rest of the culture—and ideas of animals in labs as part of the experimental apparatus is part of becoming a lab worker, a scientist, or a technician. In considering the ways in which animals have become represented in the literatures produced by and for laboratory work, we can see the tension between these two contrasting visions of animality. And those tensions are intensified by public disapproval of many animal experiments, so that it is not surprising if we find that images of animals—in the common-sense, naturalistic way that most people would think of them—disappear from scientific narrations.

Here, then, we see some of the complex dilemmas that characterize the identities of scientists, animals, and the public. The animal displayed in advertising as simultaneously both tool (or good) and "helper" reflects the dilemma scientists face in dealing with the disparate cultural conventions in which animals are variously agents and objects.[14] The issue for scientists is that they are constantly being obliged, at least in certain countries in the West, to face the agential version of the animal and at the same time need to be enculturated out of this convention.[15]

This is where the standardization of animals is key. The standardized animal is part and parcel of discursive practices which rationalize various entities, procedures, and processes, and portray them as devoid of ambiguity, uncertainty, or variance. Scientists know very well that science is not like that—it is full of uncertainty and messiness; yet scientific practices and language continue to project this image of certainty. The standardized animal thus stands at the complex junction between science and the public (and their respective relational identities) in a number of ways. For instance, as a tool the standardized animal is part of the scientific technical repertoire to which the public is not party—creating further divergence between science and public identities.

Lab workers, technicians, and research scientists alike must learn to operate within the networks of systems that constitute modern science. They must take on board the various assumptions about generalizability, need for standardization, and so on, that are crucial to the whole idea of comparing the physiology of a small rodent, say, within a lab context to that of a human in very different contexts. And they must learn to strip away these contexts to do so. One of the more significant contexts is the wider cultural understanding of animals as naturalistic, an understanding shared by people outside science; this meaning of animals must be replaced, at least in part, by a concept of the lab animal as part of the apparatus of science.

These context-stripping steps are crucial for identities, since both scientists and technicians must learn to accept the necessity of making the comparison between lab animal and clinical problems (for which the cost-benefit calculus is part of the equation), if they are to emerge as workers doing animal research. But arriving at this point is fraught; however much lab animals are represented in word or photograph as tools of the trade, students facing animals in the lab for the first time are much more likely to see them as naturalistic animals.

In this section, we have looked at what seems to be the progressive standardization of laboratory animals. We have seen how particular animals have been chosen and bred to particular standards, and have noted how this standardization has been extended to their environment in order to secure generalizable results; we have also explored how they have been represented as standardized and reliable commodities. Implicit in this is that the humans who interact with laboratory animals need to act in similarly standard ways—routine, it might seem, is what maintains standards. However, we have also noted that such standardization is not merely the outcome of breeding, genetic modification, or environmental design but also a local accomplishment in which lab workers must learn to handle animals in particular ways. We are, then, faced with dual identities of lab animals as standardized and as unique, and as separated from humans yet interacting with them. In parallel, the identity of lab workers might be characterized on the one hand by routine, and on the other, by flexible and contingent responses.

In the next section, we begin to explore the identities of the lab workers in more detail. Firstly, their view of animals as standardized—which necessarily means as objects—is not something that is automatic. It is a perception that must be inculcated and maintained. We will examine the complex processes by which such perceptions are sustained, but also variously problematized. As we shall see, such perceptions map onto a range of identities that entail various identifications with, and differentiations from, a range of actors, including colleagues, publics, and regulators. Paralleling the dichotomy we found between the standardized animal and the complex local nonstandard practices which sustain it, we find something like a "standardized lab worker" or scientific supporter of animal experimentation (standardized in that the range of arguments to support animal experimentation are limited), which is sustained by local social practices of identification and differentiation marked by complex and contradictory discourses.

PART II

ACQUIRING IDENTITIES

As we saw in the last chapter, lab animals tend to become invisible in written reports, their lives veiled. But what about the people? Human participants have to learn to do lab work, not only in the obvious sense of learning how to use particular pieces of equipment or perform certain procedures, but also in the sense that they must learn what is acceptable behavior, to take on the "body language" of science. Part of this, for many students in the biomedical sciences, involves doing things to animals that would be completely unacceptable outside science. Students must, therefore, abandon beliefs about naturalistic animals and adopt a stance of objectification if they are to succeed in their courses and become professional scientists.

In this section, we turn to people and identities in science, drawing extensively on our interviews with, and observations of, scientists and technicians working with animals, in both the U.S. and the U.K. In chapter 4, our focus is on training. Traditionally, secondary school students in biology have long been expected to dissect animals as part of their education, and educators have insisted that it is essential for students to acquire appropriate knowledge about how animal bodies are put together and function. Students resisting have not generally been supported, often finding themselves labeled squeamish or penalized in their grades. Students moving on from high school to veterinary or medical schools, moreover, must make the transition to labs using live animals. In order to survive these challenges to their moral and emotional selves, students must either drop out of that course or must find ways of coping—of justifying their actions to themselves.

In discussing these issues, we will use such terms as "coping" and "squeamishness." We use these terms in a particular way to reflect our view that these feelings are constructed and performative (e.g., Harré, 1986; Bendelow and Williams, 1998). That is, there are certain cultural expectations about how one enacts oneself in relation to naturalistic and laboratory animals; so what students must learn to deal with is the difficulty of moving from one set of expectations or norms (around pets, say) to another (around dissection, for example). Saying that these enactments are performative means that students and scientists must present themselves and their emotions in particular ways to various "audiences," including peers, colleagues, teachers, and communities. As such, these emotional responses are intimately tied up with issues of identity.

If people survive and continue in science, they must take on the social mores of the wider scientific community. But they are usually well aware of what is at stake, and of public hostility toward animal experiments; they have to find ways of coming to terms with that dilemma in the long term. In chapter 5, we examine some of the ways in which lab workers make sense of what they do, setting boundaries around what they consider acceptable (for them or for others). Research scientists and animal technicians have different roles in lab work, but are similarly affected by the dilemmas of experimentation, and we will look at some of the similarities and differences here and their implications for lab workers' identities and relationships with the animals they use.

Those relationships, however, occur in a wider context, which shapes how humans and animals interact in science. In chapter 6, we turn to some of that context, moving from (human) social divisions within science to its economic, political, and legal framework. Learning to do science means also learning to operate within these—sticking to the laws governing how scientists can use animals, for instance. However well scientists have learned to adjust to moral dilemmas, it is these wider contexts which force them to confront public opinion.

IV

Becoming a Biologist

> The anatomy lab . . . is a ritual space in which the human
> body is opened to exploration and learning, and in which
> the subjects of that learning engage in reshaping their ex-
> periential world.
>
> —Good, 1994:72

Biology is the scientific study of the living world. But not all who study living things are biologists—they may be natural historians, or lay ornithologists, for example. Generally, a person becomes a professional biologist by going through a long period of education—initially a broad training in the sciences including biology, and later, a more specific training in methods required for a working professional scientist. Like all professional training, becoming a scientist requires taking on board much more than textbook knowledge. People embarking on biological careers often begin with a deep love of nature: yet, they often find they must compromise this emotion through the very methods on which science relies. How, for instance, can someone reconcile a passion for animals with the practice of using animals in ways that might cause them suffering?

To work in the life sciences means confronting several potential moral and emotional dilemmas, such as experimenting with animals or confronting the human cadaver. As the quotation above reminds us, medical students must learn to see and experience human bodies in new ways, just as do students who work with animal bodies. For the young student in high school, the first such dilemma arises with dissection—a practice long considered to be an essential part of biological education, yet one which increasingly is being questioned. More and more students seek to opt out: sometimes, this may be an emotional "yuk" response to doing something considered "gross" or messy, although some of the resistance is due to more clearly articulated moral objections. Students may find that doing dissections is a psychologically traumatic experience, with the potential to disturb, particularly for girls (Lock, 1994).

Despite such resistance, however, biology educators continue to expect some experience with dissection. While dissection is now much less used to teach manipulative skills, it continues to be justified as a means of instilling understanding of anatomical structure. Biology teachers, and their professional organizations, argue that doing dissection is the best way to learn how bodies are put together, and to gain a sense of what tissues look like. In the early 1990s, approximately half of British school students in one survey had experience of dissection in school laboratories (Lock, 1994), with up to 80% doing so in American schools (Orlans, 1993).

Negotiating moral hurdles posed by tasks like dissection is part and parcel of the professional training required to become a life scientist. And although many students feel qualms, most do actually get down to the task and find ways of psychologically distancing themselves (Arluke and Hafferty, 1996; Barr and Herzog, 2000). Whether or not the student goes on to a career in biomedicine, he or she must distance him- or herself to carry out what is a kind of moral "dirty work"; coming to some kind of uneasy terms with such procedures thus becomes a crucial stage in the process of becoming a scientist. As the student progresses, that kind of psychological distancing becomes ingrained. Distancing, moreover, is central to the mindset of doing science, as the subjective move necessary to scientific objectivity.

Children bring to their first lab experiences a whole plethora of emotions and expectations regarding animals. In mythology and fairy stories, for example, animals are depicted in enormously complex ways, including images of hybrid figures and animals with human characteristics. From modernity, too, we inherit a tension between enlightenment reason and a will toward romanticism and sentiment (Williams, 2001), which has shaped the changing ways in which we in the West culturally perceive animals.[1] Children must learn to navigate these complicated and contradictory understandings of animals.

Young children, in the process of developing understanding of the relationship between self and other, are likely to have a sense of connection to animals, especially ones they know; but as they get older they must learn to distance themselves from animals symbolically, especially in the case of animals that are exploited (for meat, for example). Older children must, in effect, learn to achieve some degree of emotional distancing and disidentification from such animals, even if not from others (Myers, 2002). Even so, they bring to the school laboratory a set of emotional responses to animals that are immediately challenged by the demand for objective distancing—a problem made worse if they do have some identification with the animal (a cuddly rabbit, for instance). They face considerable emotional work in learning other conventions in which the animal body becomes something used for human gain. Added to that, they have to acquire what is part of the "body language" of science, an objectivity that does not permit overt emotional response.[2]

Embedded in that body language of science is a set of practices which—as we have seen—act to minimize the animal as naturalistic. Animals seen as part of the apparatus of the lab, their sequestration in separate spaces, and highly stylized forms of report writing—all have the effect of negating animals as animals. These are all part of a process of knowledge production in which the emotional responses of people to undertaking potentially unpleasant tasks are, not surprisingly, played down. And young students must learn these covert lessons just as well as they learn the layout of anatomical structures.

Dissection is one of the first major emotional hurdles students face in learning biology, and to do it, they must deploy various psychological defenses; these, as we will see, comprise a kind of denial. In this chapter, we will look at how younger students manage their experiences of dissection in the middle school classroom, and then at how older students in medical school learn to perform physiology experiments with a living, anesthetized dog. What both scenarios have in common is that while students express revulsion, many end up describing the experience in positive terms. What they have learned, we argue (apart from whatever the physiology teaches them), is a kind of absolution that permits denial of responsibility, particularly in the case of students approaching the use of living animals. From there, we will ask what role the dissected animal (or its computerized equivalent for those who opt out) plays in the relationship between scientists, students, and the wider public. Because of its significance in ethical debates and its role as a rite of passage, dissection has considerable implications for scientists' identities, which we outline at the end of the chapter.

Middle School: Learning to Dissect

What are students learning when they are asked to dissect? Dissection has been part of the curriculum for over a century. Its pedagogic purpose is to instill in students an understanding of how tissues and organs fit together within different kinds of body structure. A generation ago, students would typically be expected to dissect a range of organisms, including invertebrates such as earth-worms or large insects, and lower vertebrates such as dogfish, as well as frogs and small mammals. Whether those of us who went through the experience can remember much of, say, the cranial nerves of the dogfish is debatable; what we probably remember best are the uneasiness (in those days, less commonly expressed publicly) and the pervasive smell of formaldehyde.

Despite recent protests, dissection remains firmly embedded in curricula; it serves as a significant rite of passage, which students anticipate with a mixture of feelings. As such, students are learning much more than just the anatomy and appropriate techniques—they are also subtly learning the underlying beliefs of science, which condones the literal taking apart of the body being dissected (epitomizing scientific reductionism). They are also learning about what values are embodied in potential career choices. "When I was studying I wasn't really sure whether I wanted to be a veterinarian or not. If they had forced me to do animal experiments it just wouldn't have been worth the cost, and I would have stopped," commented one Dutch veterinarian.[3]

How, then, do students react to their first experience of classroom dissec-tion? Young students are confronting something they are likely to find morally repugnant at an age when they are developing their sense of identity and their moral character, yet are unable easily to resist figures of authority. If students find the experience traumatic, they may simply withdraw from that course or from that career path. But those who do decide to continue ultimately must find ways to dissociate their feelings: they must become desensitized. That may be a strategy essential for self-survival, but is it a strategy that is desirable? People go into, say, veterinary schools with a love for animals, and are taught that they should "do no harm"; yet the very curriculum asks them to do just that (Arluke, 2004). Meanwhile the very people who might find ways to cre-ate alternatives, the ones who questioned most, end up leaving.

Not surprisingly, most students express discomfort or anxiety about the forthcoming class, if they are interviewed prior to dissecting. They are being asked to have a physical contact with the insides of an animal—an act which for most children would constitute a clear break from many other childhood understandings of animals (especially for urban children, whose primary expe-rience of animals is with pets). Interviewing middle school students anticipat-ing a lesson based on dissecting a fetal pig, Solot and Arluke (1997) noted that nearly every student reported feelings of disgust. To get through the class, stu-dents developed a strategy to manage these feelings. But, as Solot and Arluke argue, these are not necessarily individual strategies; on the contrary, students tend to use similar means of coping because everybody around them—faculty, parents, other students, and the wider culture—provide models for how to cope and what to believe. These convey the impression that students must view the pigs as mere specimens and should not feel (or show) uneasiness.

One common way of dealing with emotional responses was for students to transform the contact in some way, to see the animal on the bench as different from their contacts with animals in their everyday lives. Partly, this involved perceiving the pig as a specimen rather than an animal. Students might begin, for example, by describing the pig as "shriveled" and brown; then, once the body is opened, familiar signs of animality, such as fur, disappear behind the newly revealed innards. Teachers and textbooks help further to narrow the focus, by directing the viewers' gaze *into* the body's interior. So students shifted their view away from the exterior, which could convey meanings of animality: one student observed that "You opened it up and the pig just like flapped down. You didn't see [the animal] when you looked at it. You didn't see the pig, you just saw like insides" (Solot and Arluke, 1997:35).

Observing the responses of slightly older students to a fetal pig dissection, Barr and Herzog (2000) noted that most do get over their initial squeamishness at touching organs, although some do not, and remain distanced and withdrawn throughout the class. And while for the majority it ended up a positive experience, perhaps through teaching them about the similarity of mammalian systems, most developed some strategies and distancing mechanisms to cope with their emotions. One way of doing so was to refer to the fact that the pigs were slated to die: "I mean, the pigs were gonna be killed anyway," asserted one student. "It didn't matter, so you might as well get a good education and make the most of their unfortunate death" (quoted in Barr and Herzog, 2000).

The younger students interviewed by Solot and Arluke also transformed the contact by referring to the pigs as "already dead." The pigs, many stated, were not killed for the purposes of dissection; rather, they reported being told by the teacher that the pigs had been "born dead."[4] It was clearly important to many of them to believe that the animals were already dead and far removed from any connection to killing. For some of these students, avoidance of dissection altogether was the means of coping. This was not a particularly easy option, as it required repeated requests by students, combined with active parental permission in the form of a letter to the school, and then meant sitting outside the classroom doing textbook activities. Teachers conveyed the impression that opting out was only marginally acceptable.

Other reactions to the emotional demands of the class were to resort to humor, or to adopt a macho attitude. Macabre humor is a standard response to first confronting a cadaver in medical school, so the use of jokes as a way of dealing with dissection is not too surprising. Nor is the gender stereotyping: everyone who commented about girls felt that girls were more likely to be squeamish or to ask not to dissect, while boys were more likely to carry body parts around the room and generally be "wild." Boys' responses were more likely to be determined by the need to perform or display masculinity (Solot and Arluke, 1997:41–42).

As they progressed with the dissections, most did not identify or empathize with the animals, but instead "adapted" to the situation. Such separating out of the emotional self from what is under investigation is both a necessary part of doing science, which must be learned in order to experiment with animals (Arluke, 1992), and is stereotypically masculine in Western cultures, as feminist writers have noted (Keller, 1985; Birke, 1994). In the classroom observations, this sometimes took the form of a rather aggressive interest in "cutting up," similar to the "aggressive cowboyism" occasionally seen in research laboratory workers (Arluke and Sanders, 1996), which would seem to be at odds with learning or maintaining much respect for life.

Once students had begun the process of transforming the animal into something else, thereby managing their emotional responses, they became more open to interpreting the experience in a more positive light. Afterwards, many spoke of it as "fun" or "interesting," as an important thing to have experienced. In their study of students in the twelfth grade, Barr and Herzog (2000) noted how many reported fascination, a sense that they got something extra out of the physical contact. Most felt that they "comprehended more because they could cut skin, feel the textures of organs, and hear bones crack" (p. 62).[5]

What is clear from these studies is that students do—indeed, must—develop emotional strategies for coping. Although opting out is in theory a possibility, it is evidently one that is not well supported by science educators. So students have to find ways of dealing with what they expect to be a difficult experience. In order to make sense of it, they must reduce their uneasiness and draw on a cultural framework within which science, nature, and animals are related in a clear hierarchy. Whatever else they may learn in dissection classes, students are learning a great deal about the authority of science, and about scientific practices as ways of understanding. And as Solot and Arluke point out, not all of what is thus learned is positive: the activity, they argue, "risks imparting to students a callous attitude toward animals, nature and the natural world, and it may dissuade some students, especially girls, from pursuing any type of science" (p. 50).

Not accepting any aspect of this practice—because of feeling squeamish, or for more overtly ethical reasons—means that such students are unlikely to opt for studying science (as one of the interviewees in Barr and Herzog's study pointed out, experiencing dissection "probably helped me decide that I didn't [want to do science]" (p. 66)). Any student wanting to do science for its own sake, or as a prerequisite to doing medicine, *must* come to terms with practices such as dissection, or—later—the use of living animals.

Medical School: Using Live Animals

By the time students enter courses that require work with animals (particularly in medical or veterinary schools), they have already made their choices and have begun to become socialized into the expectations and beliefs of the scientific community. But they face a further challenge to their sensibilities: the requirement to use living animals for instruction.[6] They know that they will be required to do things that for most people would be wholly unacceptable or at least "gross."

Cats and dogs may be used for students to learn, for example, about cardiovascular physiology; after the practical, the anesthesia is topped up and the animals thus killed. Not surprisingly, this is a laboratory practical which students anticipate with trepidation—many of them have a pet dog or cat at home, living as part of the family. Similarly, students in psychology courses can be required to do potentially cruel things to animals, such as using aversive stimuli, or food/water deprivation—practices which they may consider unjustified and unnecessary for their learning (Cunningham, 2000).

Experiencing a lab using live animals from companion species such as dogs is undoubtedly emotionally trying and desensitizing. Emotions, however, can be symptomatic of deeper concerns, such that the experience is also *morally* trying for students—reflecting a concern for presenting a moral self (Goffman, 1959). By the time medical students confront the need to do these physiology labs, they have been partly, but not fully, socialized into the culture of medicine or science. As lay actors, they draw on conceptions of animals—especially dogs and cats—as naturalistic creatures (Lynch, 1988), like pets. So the students are faced with the need to justify, in moral terms, their actions.

Like younger students about to carry out dissection, medical students before their first experience of "dog lab" consistently reported feeling apprehensive and squeamish (Arluke and Hafferty, 1996)—particularly if they thought that they would encounter a dog like their own, or one that was a breed or type they found appealing. One student elaborated,

> I had a really negative expectation. I almost didn't go. I've had a lot
> of dogs. Dogs were always real important to me. I hunted with them.
> I trained them. They were a lot of times my best friends, like my dog
> Sam. I ran with him every morning. So I felt it would be really hard for
> me to see a dog in that type of setup. . . . I agonized over it quite a bit
> the week before.

Not surprisingly, most students could not imagine using their own pets; several also emphasized that they would not want to use dogs that looked like their own animals. "I am sure that if it was like my dog at home it would matter," commented another student, ". . . it would make a difference in how much empathy I had for the dog. The thought of someone losing their dog and being out of town and their dog ends up on the tables here . . ."

Although they knew that the dogs would be deeply anesthetized before they got there, the students also knew that the dogs would be killed at the end, which was also a source of anxiety. Some expressed this in terms of their own selfishness: "I felt like killing a dog was sort of selfish in a way—you know, only for my learning this dog is going to die. I feel kind of bad about that. There's a real difference between working with a cadaver and working with something that is living" (Arluke and Hafferty, 1996:208). This presented another moral problem: they wondered how justified it is to kill purely for self-gain.

Yet whatever their misgivings, the medical students—like those in middle school—were much more positive after the event. Nearly all stressed that they had come to view the lab as an extraordinarily beneficial educational experience. This transition, suggest Arluke and Hafferty, happened because medical school culture provided absolutions to students that effectively neutralized their moral and emotional apprehension, and replaced it with a sense of fascination and awe.[7] These are not excuses or justifications for carrying out acts that are morally dubious (both of which would admit that the act is inappropriate): absolutions, in effect, deny everything.

These took the form of either evading responsibility ("someone else was ultimately responsible"), or denying wrong. One form of denying responsibility came initially from the teachers, who emphasized that these dogs came from the pound and were due to be put down anyway.[8] Students could thus take this argument on, and suggest that the dogs already had a death warrant; "it's not a big deal if the dog was going to be put to sleep anyway and we learn something out of it," one said (pp. 209–210). Some students felt that even procedures they themselves had done were not responsible for the animals' deaths—the shelter was.

To believe that the animals were as good as dead already reduced students' uneasiness because they did not feel responsible for the dogs' fate—"it wasn't like we picked them up off the street and said, 'Oh, you're such a nice big dog. You're ours.' They were already scheduled for sacrifice," commented one student (p. 210). Indeed, this rationale was held so strongly that students found it difficult to think in alternative ways. When, for example, the possibility of rescuing or adoption was raised, several students replied that "you can't take every dog home" or that there were just too many pound dogs. Just as with the middle-school students, the "already dead" argument helped to justify students' involvement.

Nearly all the students took on this rationalization. One student, however, did not, noting the distancing involved and the expectations of her teachers:

> It did die. It's not that dog would have died anyway. I mean, what are you going to do, go to the nursing home and find all of these terminally ill patients and say they're going to die anyway, so we may as well cut their hearts out or inject drugs and watch their reaction. That doesn't justify anything. You are the one doing it. This was one of the ways of stopping us from saying to ourselves—"I'm going to kill an animal". If they had just said "we are going to kill all these dogs," they would expect us to all be horrified at that.

Putting the blame on the staff was another response students made. This shift was helped by the way in which the dogs' anesthetized state was managed by the medical school. Because someone else had the job of anesthetizing the dogs beforehand, of making sure they were dead at the end of the lab, and of removing the corpses after dissection (often while the students were still in the room engaged in another task), the students never saw either the conscious animal or the completely dead one.

Because of this, students were able to think that the animal was as good as dead by the time the lab work started: "Because they put the dogs asleep for us, when we came in it was easier to deny that they were ever alive. You could think of them as dead already," commented one (p. 211). These students were taking part in a lab to learn about cardiovascular functioning; accordingly, they were required to do things to the animals' hearts which in the end were likely to cause the hearts to stop beating. But that does not mean that the exact moment of death was easy to identify, which enabled students to feel that their own actions were not responsible. As one student noted, "It dies in the end. Nobody was doing anything to it. I think that was a nice way for it to go because we didn't have to kill it per se. It just died. It was well anesthetized, so it was like it went to sleep, but not really like I caused its death" (p. 212).

This willingness to believe that they were not themselves responsible was further enhanced by the swift removal of the corpses by technicians. Students could turn away to do something with a piece of tissue they had removed from the body, and turn back to find the dog's body had vanished. This happened so quickly that students did not have time to mull over what had transpired or to achieve emotional closure, as one student recalled:

> One thing I wanted to do was to say goodbye to the dog after we were done. We ended up taking out the heart, and we took it aside and cut it open so we could see everything inside. We were by the sink and then were done. I turned around to the table to say goodbye to the dog and it was missing. It was horrible. I'll never get to say goodbye. I think it would have helped me feel a lot better about this if I had said goodbye to the animal.

Clearly, if students are expected to take part in these labs, and to carry out acts which in other contexts would be morally suspect, then they must find ways of dealing with their emotional and moral repugnance. Denial, through shifting the blame onto someone else or onto an institution like an animal shelter, is a powerful psychological way of doing so, and provides an absolution. But the structure of the lab experience itself facilitated such psychological transition. Apart from the division of labor and the structuring of lab space, which enabled separation of students from the dogs as animals, the inclusion of the lab manual specifying clear beginning and ending of the lab and discrete bits of physiological data also helped to gloss over student awareness of more broadly defined endings, such as death. Moreover, it certainly did not encourage them to accept any responsibility for their acts in the lab.

Another strategy for coping was to deny wrongdoing. Usually this was on the grounds that the animals were deeply anesthetized and could feel no pain. This move enabled students to see the dogs as not quite real, as not quite alive, rather than simply unconscious. It also distinguished the students' actions from the laboratory "torture" described by animal rights activists; it was, students felt, "no big deal" to use them as long as the dogs were completely unconscious.

Suggesting that these animals "were not really like dogs" was another ploy for denying wrong. If students had felt that the animals were like their dogs at home, they would not have been able to get through it; what they did talk about, however, was how the dogs—lying on their backs and intubated—looked different, making them not quite dogs. Nor were they dogs in the sense that relationships with them were impossible: "When it's laying on the table on its back strapped up, its belly shaved, it's asleep, it's not as much of a dog" (p. 215).

These animals were, moreover, somewhat less than dogs in the sense that, having come from the pound, they were presumed to be socially unattractive and undesirable (and hence slated for euthanasia). Students could, therefore, argue that it would not have seemed right to use animals bred specifically for the lab, but it was acceptable if the animals were already scheduled to die. By classifying these dogs as "unwanted," the students were able to rationalize their use—and deaths—in the lab. Indeed, by using them to gain knowledge, the students felt, the animals' lives were retrospectively "more worthwhile" than if they had just been euthanized. As one student commented, "The fact that these animals were scheduled to be sacrificed anyway made me feel that maybe this was okay. . . . I think it was worth it because the dogs were going to die anyway. It would not have been worth it, even for the amount we learned, if these animals had been raised just for this to have been done" (p. 216).

The students, then, learned very well how to absolve themselves from responsibility and wrongdoing. Where before the lab there had been much trepidation and doubt, afterwards there was fascination and enthusiasm—just as several younger students reported after doing their first dissections. But for these older students, there is usually no opportunity to opt out—they have already chosen their professional path and have to get through the experiences. For them, fascination is a near-universal response. They agreed, note Arluke and Hafferty, ". . . with a consistency that rivaled a ritualistic chorus—that lab dogs were far better educational tools than were books or human cadavers" (Arluke and Hafferty, 1996:217).

It was not just observing the workings of a living body, but having the hands-on contact with it that mattered here; this allowed them to apprehend immediately, most felt, whereas other methods of knowing required them "to think." One student commented, "This was really neat—digging in and seeing the stuff actually working and pumping—it was great." This was "better than books." As another student pointed out, seeing things illustrated by models was less useful because they were never living, and because the students themselves did not change anything: "To me a model wouldn't be much different than a book. Nothing you do is going to change anything with the model."

Direct contact with a beating heart did not need to be interpreted in the same intellectual way as reading a graph—rather, it provides a kind of tacit knowledge. There is an irony here, in that distancing oneself psychologically enables ways of corporeally engaging with the animal's bodily material, allowing a way of developing tacit skills considered necessary to scientific investigation.[9] By divesting animal flesh of its emotional significance, arguably the student can better (literally) come to grips with it as an object of study, thereby teaching one's *own* body certain skills. Needless to say, not all students are successful, or physically equipped to do this (see Knorr-Cetina, 1999).

The Place of the Dissected Animal: Cyberfrogs and Meanings

Some students, however, continue to have doubts, asking "tough" questions about rights and wrongs and the place of the dissected animal in their education. Many students, especially in non-medical programs, now demand that they can opt out of dissecting real animals, and be provided with alternatives wherever possible, arguing that there is little educational value for them to engage with animal flesh. Whatever qualms they might have about facing a real animal for dissection in the lab, few would hesitate about learning anatomy and physiology by engaging with the computer. Interactive computer simulations to dissection are now readily available, allowing students to carry out virtual dissections of various kinds of animals—Stanford University, for instance, is producing a range of "Virtual Creatures" for students to explore anatomically.

Such simulations have been developed and marketed to meet the growing demand for alternatives to dissection—indeed, the marketing strategies of companies producing simulations often make explicit reference to environmental concerns or to animal rights (Fleischmann, 2003). The software enables students at the click of a mouse to remove the animal's skin, or to zoom into the gastrointestinal system. As advertising for dissection simulation software typically emphasizes, it is more eco-friendly to use virtual frogs, and we can forget the old memories of foul smells and revolting mess. It might also be easier for students to learn by following carefully thought-out computer programs than following instructions on how to cut into the animal in particular ways. As one student commented,

> The lab kits consisted of third to fifth generation photocopies of pig body parts, making it nearly impossible to actually identify anything on the pig. The teacher was constantly answering our questions and frustrated as to why we couldn't find anything she asked us to. Meanwhile, the students with the alternate assignments worked with crystal-clear pictures online, took however much time they needed to copy diagrams and descriptions, and got much higher scores than the rest of us. I wondered why we all didn't go to the computers and work with clear pictures of the fetuses, instead of pigs mangled by faulty cuts or covered in . . . whatever . . . or disintegrating after a week's wear. It was useless, frustrating and demoralizing. The only way I passed my pig test was through the "Virtual Pig Dissection" website.[10]

So in terms of getting the grades, using the simulations might, for some students at least, be advantageous. On the other hand, cyber-dissection or other alternatives do not (yet) teach students the kind of corporeal skills that have long typified biology and how it is taught. You simply cannot gain the tacit knowledge of what animal flesh looks and feels like from a computer—as teachers usually emphasize. Those educators who believe that this is a crucial educational experience have, however, themselves been part of the same processes of socialization that we have described here, and may be less willing to consider the benefits of computer technology.[11]

Analyzing the relationship between the developers of computer simulations and animal advocacy, Fleischmann points out that "cyberfrogs" are examples of a value-based technology with strong connections to a specific social movement; they are often marketed in explicit recognition of animal rights issues, and their use is actively encouraged by several animal advocacy organizations. As such they are "boundary objects" linking information technology to animal advocacy. This claim draws on previous work in science and technology studies, detailing boundary objects as things which are used by two disparate social groups to create meaning (Star and Greisemer, 1989). Although that meaning might depend on the character of that group's particular social world, the common use of the object to signify can promote coordination (or conflict).

Yet if cyberfrogs are boundary objects articulating an intersection between IT and animal rights, the figure of the animal set out in the classroom for dissection is also a powerful boundary object. As a once-living animal, it is clearly linked to social worlds in which animals are active participants; as a body for dissection, it is linked to the worlds of science educators and scientific research through low-tech artifacts such as scalpels and dissection pins. It also stands between the physical space of the lab, where actual dissections usually take place, and the virtual space of the computer, which is often in another physical place.

Dissected animals, moreover, serve as boundary objects between a particular version of what counts as science and a particular version of the public. Thus, as objects of training and knowledge, they stand at the boundary between a scientistic science and a scientistic public; yet, as subjects of suffering and instrumentalism they articulate the boundary between a science more sympathetic to animals and environmentalism, and a like public. This is one reason why the figure of the dissected animal is a potent one, symbolizing both public disquiet and the scientific knowledge that the student wants to obtain. Thus, the dissected animal is deeply tied up with the processes—which nearly all biologists have gone through—of finding emotional solutions, and absolving the self from responsibility for death or suffering. Like "the" laboratory animal, whose meanings we explored in earlier chapters, the animal-for-dissection has multiple resonances. It has particular significance, as we have seen here, in students' journey through biological education and hence for their later identities in relation to science.

Dissecting Identities

However much trepidation students feel when approaching these rites of passage, they must either find ways of dealing with their anxiety or they must opt out. Because of the professional expectations of science, educators rarely present alternative ways of managing emotions, precisely because science is taught as though it is devoid of them. Often, as we have seen, opting out is far from easy, and only the more persistent students will follow that path if dissection is a course requirement or is widely expected. Teachers tend to take a dim view of students opting out, arguing that they will have a worse education, and perceiving "squeamishness" as an inappropriate emotional response.[12]

Students who do not easily come to terms with morally or aesthetically distasteful procedures then have two options: if alternatives are available, they may take those (and deal with the possible opprobrium of staff and other students). Or they might simply make the choice not to continue with studying biological sciences or science in general. Not only is dissection a crucial stage in becoming or not becoming a scientist, how people respond to it then affects how they identify in relation to science. It is quite difficult to develop an identity as a biomedical scientist and yet still reject these educational passageways into the career.

As we have seen, students deploy various methods for dealing with emotional unease—believing that the animal was already dead, that someone else was responsible, or denying that they were doing wrong. If they want to pursue careers in biology, these psychological defenses must become second nature; more to the point here, they must become part of the person's identity. To attain a sense of oneself as a scientist, pursuing objectivity, learning to distance oneself, is a crucial step.

Identities, as we noted earlier, can be conceptualized as emerging out of engagement within particular networks (Michael, 1996), most notably in this case those of other scientists or science educators. It is these networks which help to establish boundaries of what is acceptable by "persuading" newcomers to accept particular "facts" and "values." For example, in the process of entering a biology lab class, students are being taught the value of physical engagement with biological materials—it is in this way that they derive an understanding of the complexity and recalcitrance of animal bodies (and other materials, such as plants) that cannot be captured by representations (e.g., diagrams in books). Through such physical engagement one learns to "see" bodies "properly."[13]

As students move from revulsion to fascination, they are learning to deny their responsibility for their acts and their wrongfulness. Arluke and Hafferty (1996) point out that students are learning a kind of motive talk—that is, ways of speaking that recast the meaning of behavior when people have broken social convention. Students know that they are breaking with wider convention, but must find ways of presenting their motives. The kind of absolutions students learn may be made easier in relation to activities such as lab work which largely take place out of public view; in such situations, lab workers are usually addressing their peer group rather than a wider public.

Learning "motive talk" has implications for professional identity. "Although dog lab is but a brief experience in the students' larger medical education," Arluke and Hafferty argue, "it can serve as a powerful reminder that technical skills can be sharpened only by quelling or suspending moral doubts . . . they learn that it is acceptable, even necessary, to suspend asking 'tough' questions in order to get on with their 'real' learning, which they do with a sense of excitement and awe rather than moral trepidation" (p. 223). In other words, they learn to "talk the talk" as part of becoming socialized into medical/scientific communities.

Most people emerge out of the dilemmas of scientific training with a kind of uneasy truce. They appear broadly to accept that animals will be used—or to use animals themselves—but they retain some of their earlier ambivalence. This in turn has implications for their relationship to wider networks. Some researchers or technicians working with lab animals, for instance, prefer never to discuss their work with anyone outside of the scientific community; if asked, they might phrase their answer ambiguously. So while this might involve a clear identity as a scientist within those networks, that identity may be rather less clear outside them.

As we will see in the next chapter, many scientists working with animals are uneasy acceptors—they may emphasize their acceptance of animal-related practices but at the same time stress various ways in which they draw the line—much in the same way that students comment on how they couldn't use dogs of particular breeds or animals that they thought had been bred for dissection. In interviews, scientists tended to present themselves through a kind of rational emotionality—that is, they defended animal use, including their own, but acknowledged past and present emotional reactions to it in the form of admitting that there were some things they could not do (see next chapter). By expressing ambivalence, these scientists are indicating their engagement in quite different networks, allowing them to claim identities as both scientist and simultaneously as respecter of animals.

V

The Division of Emotional Labor

> My work . . . is to act as a buffer between the
> animal and the scientist . . . to be the first line
> of defence for the animal.
> —Animal technician, U.K.

In this chapter and the one following, we look at how people working in the lab make sense of what they do. In part, this has to do with making sense of particular experimental procedures that may involve pain or distress to the animal, prompting researchers to find ways of distancing themselves. But in part, working in the lab also has to do with taking care of animals, as the opening quotation implies. This means having a relationship with them that is much closer to the relationships people have with companion animals outside the lab—as well as having to do with relationships between different lab personnel.

Who's Who in the Lab

Laboratory experiments using animals involve many different people; experiments may be carried out in such institutions as universities, private research foundations, hospitals, animal breeding facilities, or pharmaceutical companies. Broadly, however, there are several kinds of workers directly linked to animals. Carrying out a research program can require a principal investigator, doctoral students, research technicians, and animal caretakers, as well as various others. We have not, however, directly studied inspectors who oversee standards of care or named veterinarians who oversee lab animal health and welfare (see Carbone, 2004). Rather, our studies have concentrated on those people who have a direct, day-to-day interest in the research—the lab scientists and technicians/animal caretakers.

In the last chapter, we considered how people come to terms with the emotional dilemmas of using animals for dissection during the course of their training in biomedical disciplines. If students believe they must go through these conflicts to gain the knowledge required of them, they will find ways of coping. By the time someone gets to the stage of doing doctoral research, most have come to terms with these difficult choices—or have opted either to get out of biology or to specialize in areas posing fewer dilemmas. They have made a decision to enter biomedicine professionally, and usually express a desire to solve biomedical problems or to help alleviate human (or nonhuman) disease. If they are to do research projects based on animal use, people must accept certain procedures (which may be invasive and potentially painful for the animal) and they must accept that keeping laboratory animal stock in large numbers necessarily entails culling surplus animals. Accepting procedures does not mean that each person necessarily has to do them, but means accepting that they are part of a research *process* that can entail pain. As we will see in this chapter, scientists quite often express a dislike of some experimental procedures while accepting others (as less invasive, or as less unpleasant for either researcher or animal); what this means is that they personally draw the line at doing some things, even while knowing that the same things are carried out regularly in other places or by other people.

Most institutions in Europe and the United States now require, as part of the PhD process, that students are formally trained in a number of laboratory methods and forms of analyses. This can include specific training in handling laboratory animals. Doing a research degree is also a kind of apprenticeship, where people acquire skills in experimental design and lab practice by carrying out experiments alongside other, more experienced scientists. After the research degree and subsequent short-term research posts, people might move on to more permanent jobs, and into the position of becoming principal investigator. In this role, they will usually seek funding from granting agencies, run their labs and oversee their management, and liaise with institutions. At this stage of their careers, some PIs are no longer very active in the day-to-day running of experiments, and hence have less contact with laboratory animals. Nevertheless, they are the ones who are *responsible* for the projects, and who must therefore find a rationale for them—arguing the potential medical benefits, for example, as a reason for funding, or perhaps acting as spokespersons for the research community.

Science does not get done without technicians working alongside research scientists. Among these are the animal technicians, or caretakers, whose primary role is the maintenance of animal colonies and the daily care of the animals, as well as more specific care of animals postoperatively. They must also deal with the purchase of lab animals from external breeders, food and bedding, and ancillary equipment such as cages, cage washing machines, and so on. Like research scientists, animal technicians must learn to accept some aspects of the job, such as surgical procedures, tumor development, and killing animals. Unlike research scientists, animal technicians are not always graduates, and they sometimes enter the job principally because they want to work with animals rather than out of a desire to do experiments.

In addition, the work cannot be carried out without staff in supporting roles—veterinarians, inspectors (to fulfill legislative requirements for human health and/or animal welfare), members of ethical committees, and so on—all of whom can have a direct bearing on the conduct of laboratory workers and animals. In Britain, for example, institutions and researchers must be licensed to carry out procedures with animals, and must liaise with a government-appointed inspectorate. The inspectors' job is to monitor research projects as well as environmental conditions of animal houses, so that animals are well looked-after within the parameters of legislation. Because inspectors can require alteration of project proposals (to reduce animal numbers, say) and can expect changes to animal husbandry in animal houses, the work of both research scientists and animal technicians is directly affected by their decisions.

Shared Coping Skills

In some ways, scientific training instills in students an ability to distance themselves, to be "objective," which is at times at odds with their relationship to animals in other contexts. They must, as we saw earlier, acquire the rhetoric through which animals become data. Many training courses in lab procedure will do just that, reinforcing the emphasis on the technology and what it produces. But some training is more ambiguous, allowing animals still to be seen as naturalistic animals. Courses in animal handling obviously do so, since good handling is not only about reducing animal stress, but also about avoidance of injury to the handler. Learning to avoid biohazards and the risk of transmitting disease between human and lab animal also means recognizing the animal as animal. All personnel using lab animals encounter this schism and must learn to negotiate between both senses of "animal."

Because of this ambiguity, professional socialization for lab workers entails learning to switch between objectification (keeping animals at a distance) and identification with them. In this sense, they must deal with ambiguity on a daily basis. Sometimes, lab workers do find themselves identifying with particular animals; more often, they find ways of setting emotional limits. While objectifying is a necessary part of taking the role of lab scientist, it is not necessarily easy or part of the self. For some, especially some animal technicians, distancing feels unwanted or unwelcome, seeming to contradict their desire to work with animals as living things.

Some ways to deal with the ambivalence of using animals are shared by everyone involved directly in animal research—research scientists and animal technicians alike. But they also respond to the dilemmas in rather different ways, reflecting occupational divisions. Here, we turn to these similarities and differences, drawing on observations and interviews with laboratory staff explaining their attitudes and relationships with lab animals.

Research scientists (from doctoral students to principal investigators) and animal technicians express similar reservations about particular species, or about specific techniques. Although these are quite different, the reservations expressed indicate lab workers' ambivalence about using animals and inflicting suffering. Lab workers of all kinds also share a similar problem in identifying their occupation outside of the lab, often feeling that they must deny what they do or are because of public opinion.

Most lab workers we interviewed made reference to drawing a species line somewhere, whether that was around vertebrates versus invertebrates, or some species of mammals (see also Koski, 1988). One technician, for example, was at pains to explain how she would find it more difficult to work on species other than rats:

> It's not that I don't think that rats are any more or less important than say cats or monkeys, but I treat them all with the same respect, whatever, mice, rats, again I think that's a misconception of the general public. I don't see why they should think that cats are anymore special than monkeys or rats quite frankly. You know . . . I like cats, I couldn't work with them. . . . But I can work with rats. I know it seems, somehow it just doesn't seem justified, but I have cats at home. I think emotionally I would get too involved with cats. I don't think I could treat cats in the same way really. I mean, I like rats, I'll have rats as pets, they're great.

In this extract, the technician is very clearly expressing her ambivalence: she first stresses that mammals are similar, but then draws the line at cats, which she has as pets. She also points out that she feels fine about using rats, then goes on to say that she could have rats as pets because "they're great."

Because of public opposition, scientists are encouraged to seek "simpler" organisms if possible, such as invertebrates or lower vertebrates. This might also make it easier for the researcher, as one scientist explained:

> It sounds stupid, but it's different working with mice than goldfish. I'm sure goldfish don't evoke the same kind of response. I am a lot more aware of [mice] and a lot more careful and compassionate, if you like, towards them, as an animal rather than just a thing that you do your experiment on.

For him, then, it was easier to use goldfish, while for others using rodents is easy while primates are not. Another scientist explained why he could not carry out invasive work on the nervous system of monkeys but had few qualms about working with mice:

> I guess the reason why I don't like the idea is that I've worked with monkeys for so long now that I see them as a very different species to mice. I have no empathy with mice. There's no identification. They're all white, anonymous. The monkeys, you get to know them as individuals.

Similarly, asked about the moral balancing act needed to justify particular experiments, one senior scientist noted his own inconsistency, saying that "there are some species that I simply would not choose to work on. No matter how brilliant I thought the experiment was, I can't see myself doing experiments on dogs or cats or monkeys. I probably accept that someone has to do it, but I don't fancy it being me."

Research workers also can place restrictions on doing particular techniques (admittedly, this is easier to do for senior research scientists, who have greater autonomy). One scientist referred, for instance, to techniques that "make me cringe. They take cerebral spinal fluid out of the back of the neck and the thought of putting the needle through the spinal cord makes me cringe, so I wouldn't do that." Others might draw the line on procedures such as taking a blood sample from behind the eye, or using animals in long-term studies. One technician explained her feelings about a colleague's work in another lab:

> They do bone grafting there and work with the same dogs for weeks. She has to walk these dogs back and forth every morning. And she gets attached to them. . . . You can tell it really bothers her some days. . . . I could never do that.

Drawing the line at certain species or techniques, then, was something commonly expressed in interviews (and has been similarly expressed in questionnaire studies; e.g., Paul, 1995). Clearly, believing that it is acceptable in principle to use animals in experiments is not straightforward—for most people, it is in practice acceptable only for some species, some of the time.

A second shared coping skill was for lab workers to see themselves as the "good guys" (a theme explored further in chapter 7). Nearly everyone claimed to have seen or heard of abuses; but nearly everyone also expressed the opinion that animals were well cared-for in their lab. The abuses happened elsewhere, they insisted—under previous legislation, or were perpetrated by outsiders (such as foreigners, or by scientists working in other countries). There was little distinction between senior scientists and technicians in this regard.

Sometimes, scientists spoke of their experiences in labs in other countries. One senior researcher working in Britain took pains to point out that lab practice in Israel, where he had done postdoctoral work and which had no animal protection legislation at the time, was similar to practices in the U.K., which has strong legislation. Despite concluding that scientists' self-control limited abuses, he went on to demonize scientific practice in the U.S., referring three times to papers published from U.S. labs describing experiments that, he claimed, would never be permitted in the U.K.

Several respondents were more explicit about their lab, or nationality, being the "good guys." Explaining that she noticed "certain cultures are very different and they do not have the same respect for animals as we do," one technician went on to derogate "orientals" as being less likely to respect animals, and unable "to understand why we are so sensitive." Later, she went on to say that "the fact that we're English . . . is that we do tend to have a more sympathetic approach to animals anyway, I think it tends to be an inbuilt thing."

Although "foreigners" were cited quite often in the British interviews, several respondents also mentioned the cosmetics industry, which, according to one researcher, "gives experimentation a bad name." But the general public were also sometimes "bad guys," in terms of double standards and—particularly—the way some pets or agricultural animals are treated. In each of these cases, there is a clear implication that the speaker's own practices are somehow morally better than those of others outside of the lab (Michael and Birke, 1994).

Humor served as a third common coping technique. The sick joke about the cadaver is a staple of medical school. Lab personnel also use humor to cope with the dilemma of using animals. For example, most animal houses have animal-related cartoons pasted on the walls (Gary Larson is a favorite cartoonist). Sick jokes sometimes appear on the doors of refrigerators containing animal parts. Joking statements about particular animals may also be posted outside cages, along with photographs of lab animals posed in ways intended to be humorous and affectionate. In one study of the long-term effects of a life-support system implanted in pigs, technicians were required to stay with the animals twenty-four hours a day: they became "pig-sitters." This, inevitably, meant a closer relationship to the animals, bringing with it uneasiness. On the wall, the technicians displayed a series of photographs, some of the pigs dressed up in hats and sunglasses; another notice scored the pigs' "sex appeal" and "charm" (Arluke, 1988).

Since research scientists and technicians all feel they are in the same boat, they express the dilemmas of doing animal experimentation in similar ways. What they share here is a mutually created understanding, and a common sense of being besieged by outsiders' hostility. That is, by working in labs together, they create shared meanings, such as a sense of what species might not be used, what techniques are appropriate, and so on.[1]

The Technicians' Burden

Research scientists and technicians, however, occupy quite different locations in the laboratory division of labor, which is expressed in several ways. In that sense, they are not so much united by shared opposition to the outside world as divided by their different positions in power structures, and their relationships with lab animals. Technicians, for example, are more likely to separate out specific animals as "pets"; they also have to do most of the routine culling of animals in the animal house. Technicians, moreover, are likely to see themselves as buffers between the animals and the demands of scientists. These differences are not, of course, absolute: some research scientists, for instance, take their share of the routine culling. But we want to draw out here some differences in degree or emphasis.

The first of these is the requirement that animal houses routinely kill animals. Very few scientists made reference to the "background work" of the animal house—least of all the inevitable requirement that stock numbers are kept constant. To produce animals for experiments, there will be surpluses and those animals will be killed, usually by the technicians. Scientists, by contrast, sometimes dispatch animals at the end of experiments.

Indeed, relatively few research scientists mentioned killing at all—whereas most of the technicians did. One scientist, working in cancer research, commented, "I actually find it quite difficult working with any animals. I don't like or enjoy it. I don't think anyone does. I think it would actually be quite revolting if people enjoyed essentially killing animals, which is what it boils down to"—referring to the fact that all animals, once they have finished an experiment, must be killed. Another mentioned the need to kill, commenting that it was acceptable as long as there was no suffering.[2]

These, however, were exceptions; most scientists described the problem they were addressing (working on genetic disease, or cancer research, for example), or spoke about where they would draw the line on species or procedures, rather than referring to death. This fits with the use of euphemisms like "sacrifice" in written reports, described in chapter 3. Although none of our interviewees expressed it verbally in such terms (nor would we expect them to), there is much about the use of animals in laboratories that mirrors traditional ritual sacrifice (Lynch, 1985; Arluke, 1988). Just as the victim of traditional sacrifice takes on symbolic significance, so too does the lab animal in the process of becoming data. A proper scientific sacrifice is one where death enables the animal to be linked to the larger purposes of the experiment; put another way, animals must not be sacrificed in ways that jeopardize their experimental value. One university technician put this clearly:

> . . . a room [in the animal house] I look after is an infected room and they deal with lungs and they're doing like a flu and we do get a few dead mice in there which the scientists don't want to see . . . afterwards they want to take out the lungs of the live ones. I've watched them do it and that doesn't bother me at all because I see them using it for something positive. They're not just wasting animals, they're actually using them for a purpose.

She draws attention first to scientists' not wanting to see the dead bodies, and second, to animals that died in their cages and so cannot become part of the experiment (even though they probably died because of the introduced infection), in contrast to those animals which are sacrificed for their lungs. Then, and only then, is their death justified as "something useful."

Because of the symbolic significance of such killing, new recruits to the lab must learn to think of their first sacrifice as a rite of passage—just as students must when confronting the first dissection (Arluke, 1988). What matters here is learning to sacrifice animals *in order to* collect tissue or data: the precise methods of doing so must be learned. On the other hand, if an animal is still alive but the scientific procedures have been finished, its death is not crucially part of the generation of data, and it matters less who does it. As Arluke points out, it is the results of the experiment and not the animal per se that are sacred in the scientific rituals of sacrifice.

Technicians too must learn the language of sacrifice, as one commented:

> Some people are more comfortable saying "sacrifice." I think it is a way of

announcing that you are killing them . . . yes, you know you are killing them. "Sacrifice" is just a word you pick up. I don't think it helps you justify the experiment in your mind. The purpose of the experiment is what justifies it, not having a word.

Another referred to animals that are "killed off for the results to be brought about." He felt that one of the most difficult things was when animals came back to the animal house after surgical procedures and were "obviously suffering"; if an animal is taken away and then killed (by the scientist) that was less difficult, he said.

One aspect of killing mentioned by nearly all technicians (but hardly mentioned at all by research scientists) was the need to cull. This was often expressed as a gripe by technicians, complaining about scientists' lack of adequate planning, with the result that the animal house overproduced stock, which would then have to be killed. There seemed to be two concerns here: first, that technicians would have to do this routine killing; and second, that these surplus animals had been bred for no purpose—"we've just bred them and they're no use at all," said one technician, expressing her concerns about the need to cull. Another put this dilemma in broader terms:

> . . . like all of us . . . we're born and we die. Now the question is how soon is it before we die. Now as far as culling animals is concerned if they weren't culled as a litter, which really is the best time to cull them, then of course they're going to be killed off later on because nobody will use them. There will be no need for them so what do you do? You can't just continually keep hundreds and hundreds of animals that nobody wants.

There is a reluctant acceptance here, that animals must be culled and that perhaps it is better (arguably easier for the person doing it) if the litters are very young than having to kill older animals. But the extract also alludes to the problem of overproduction, which was mentioned by many technicians. At times, they mentioned that killing surplus animals was more upsetting than when animals died in the course of experimental procedures—the latter they saw as being some use so that the animals' deaths had a purpose. At other times, however, they explicitly criticized the scientists for failing to plan, with the result that animals were often bred but then not wanted:

> I totally disagree with it, animals just being bred up and there is no controls of what kinds of numbers people wanted. You might get 200 rats a month which nobody really wanted and they were destroyed and really there was no reason. . . . It's difficult but I think people should be told to make sure they know what they want six months in advance so we can get roughly the right numbers bred up.

On the other hand, some kinds of animals are more expensive—dogs or primates, for example—so such overproduction is less easy. In that case, technicians may have to take on the role of "scavengers"—going to other labs to see if they can obtain body parts from newly dead animals in another lab (Arluke, 1988). Or they might quietly exert control and try to coordinate the scientists: one technician working in a medical school and taking care of the ferrets used in experiments did just this: "So what I would do is, I'd get everybody I knew to say, they're having a session on X day, if you want tissue this month or this quarter, you get it from those animals, and either you store it or you adapt your procedure." So although some research scientists do cull, it is largely the technicians who are expected to carry out this unpleasant task. On the other hand, research scientists more often have to kill animals directly or indirectly as part of the procedure.

A second difference involves pain and the justification for using animals. A crucial question underlying the use of animals in experiments is whether or not they suffer pain. Much of the justification for using animals relies on the assumption that they are not suffering (because researchers believe the experiment is not painful, or that the species does not feel pain), or that any suffering has been minimized by, say, the use of anesthetics. Any residual suffering is then held to be "necessary" for the pursuit of medical benefit. Thus, one researcher in an anatomy department said that it was easier to do procedures on fish than on mammals, "because the animal obviously wasn't suffering . . . as far as I was aware—and it is possible to be aware about the sensibilities of a goldfish—it wasn't in any pain."[3]

In a study of scientists' perception of animal pain, Phillips (1993) noted that, although anesthesia was routinely used, analgesia was not.[4] Most researchers did not use post-operative pain relief, and generally did not take the issue seriously; yet the same researchers will seek to prevent infection through the routine use of antibiotics. The failure to ensure pain relief was not, Phillips argued, because of some belief that animals simply don't feel pain; rather, it is that "Laboratory animals are categorized and perceived as distinctive creatures whose purpose and meaning is constituted by their role as bearers of scientific data." Similarly, the research scientists in our interviews seldom mentioned pain, except to imply that animals did not suffer pain unduly after operative procedures: "An animal that can't speak can't tell you, you just have to assess from their behavior and we watch their behavior and they don't need [analgesics]. I mean they behave completely normally," argued one scientist, whose research involved severing sensory nerves in infant rats. Another felt that if a rat he used was obviously gaining weight after surgery it was a "perfectly happy rat"—something which he could tell, he said, because the animal had normal levels of corticosteroids (stress-related hormones).

Technicians, however, were more ambivalent. On the one hand, they painted a picture of their labs as generally well-run places where pain and suffering were always minimized. Like research scientists, they sometimes articulated the belief that no one likes doing experiments on animals, but it is necessary for medical progress. On the other hand, technicians often saw themselves as buffers between the scientists and the animals. Their role, as they saw it, was not so much to carry out the experiments as such (though some did participate directly in experimental work or in removing tissues), as it was to *care* for animals.

In describing their role as buffers, the technicians often drew attention to their attempts to challenge scientists if, in the technician's eyes, the animal was in pain. As one technician explained,

> If I think that the animals are in any pain or discomfort, even though we are supposed to minimize it and do things, occasionally it does happen more by accident than design that worries me . . . [if that happens] usually I challenge the person doing the experiment . . . if I thought the animal was in pain there was nothing I could do to help it or get anybody in fairly quickly to help it. I was able to cull without anybody really challenging me.

Another felt that "I'm a go-between between the animal and the user, so where the animal can't speak I speak for it . . . if there's a mouse that I feel needs to be killed then it will be killed, that's because [the scientists] know me well enough to accept what I say."

The buffering metaphor, alluded to by many of the technicians, not only shows work-related tensions between technicians and research scientists, but also suggests that technicians are more willing to admit that lab animals suffer pain. Relatedly, several technicians mentioned that, as time went by in the job, their conscience was pricking them more; very few of the research scientists said anything like this (and when they did, it was junior scientists who expressed it). Technicians saw it as part of their job as animal caretakers to monitor such suffering closely and to do something about it. In that sense, the animal for them is not solely a bearer of data; it remains at least partly a naturalistic animal.

Technicians were also critical of the numbers of animals used in labs. Opponents of animal research frequently argue that there are far too many animals used, and scientists are usually encouraged to reduce the numbers. Yet, several technicians felt there was considerable wastage, too many animals and (to them) unnecessary experiments. Commenting on surgical research, one said she thought the experiments were ". . . a bit irrelevant. They seemed to have done the operation, experiment, before and they'd got the results and to me they seemed to be repeating and it seems to be an awful waste of animals."

Another technician acknowledged how repetition was intrinsic to doing science, but felt that doing so used too many animals. Referring to the lab's antigen work she said:

> It might be repeated using something slightly different next time to see whether that had any difference. Or they might decide that the antigen wasn't prepared in such a way to produce the response they wanted, so then it's repeated with more animals . . . science is such that is the way you do an experiment . . . you do repeat it to confirm the results and you need statistically a certain number of animals . . . [but] I find that hard to justify.

That repeats may be unnecessary or irrelevant was a common theme for technicians. Replicability is, in principle, central to scientific objectivity: yet most experimental work is rarely replicated, or at least rarely published (Collins, 1981). However, technicians do not consider replication to be so necessary.

A third difference stems from the role given occasionally by technicians to lab animals as pets and friends. Although most animals in laboratories are identified by numbers, a few acquire names—a process which makes it emotionally harder for lab workers to conduct procedures. An even more select few are separated out from their peers, and are kept as pets. That is, they are singled out for special affection, possibly for special living conditions—and, crucially, are usually spared the death that awaits their fellows. This is common practice, even though it is in violation of existing legislation.

If these pets die or are killed, they are sometimes mourned in ways that animals remaining as numbers are not. These select animals, then, acquire a quite different moral status from the remaining animals—they are treated as sentient rather than as bearers of data, as "a living entity rather than as a container housing tissues" (Arluke, 1988:106). Arluke recounts how technicians in one lab hid a rabbit who had become a pet (and whom they had called Fat Cheeks). When the time came for him to be culled, the technicians found a way of sabotaging the procedure and keeping him alive;[5] he lived as a lab pet for another year.

Animals that become pets often have some special quality that sets them apart. In the case of one rat, that was size: the technician who looked after him explained that he "weighed two kilos and that was why he was so special and he was so tame." Numbered originally 007, he inevitably acquired the name James Bond and was a popular figure in that laboratory—"people used to come in and see James from the whole of the department," she recalled.

Another technician recalled her days working as a veterinary nurse in a small animal practice, an experience which, she said, had "hardened" her because of the callous treatment of pets by some owners. "You do get slightly hard," she went on,

> and . . . you think of the stock animals as numbers rather than actual animals, whereas we have pets in the cage wash [area] and if something happened to them I should be really upset because they are our pets and we fuss them and they sit on your shoulder and they're like pets. Whereas the amount of mice in one room in the cages you don't actually associate them as being pets or animals.

Pets here are clearly singled out from the numbered mass of stock animals and spared the fate of the others.

Not surprisingly, research scientists very rarely referred to animals in the lab as pets (although this does not mean that they did not treat some animals differently; rather it may mean that they were less willing to admit special treatment). Nor did they refer specifically to their *relationship* with any of the animals. The technicians, by contrast, were keen to emphasize their friendships with the animals in their care. Several spoke of how they said "hello" to their animals when they went to work in the morning or put the radio on for the rabbits.

Arluke (1990) suggests that technicians share with their charges the fact that both "work" for others to make the experiments possible. "Sustaining pet-like feelings towards selected laboratory animals," he notes, could create ". . . a limited but genuine sense of solidarity between humans and animals" (p. 200). Whether or not a person bonds with a particular animal obviously depends on many things—the person, the animal, the situation. Certain animals are not granted special status: one researcher in a dog lab felt that "when you get a bunch of greyhounds in the lab, they just sort of stand there and look strange. You don't feel as badly anesthetizing them as you do a little frisky cocker spaniel that's around looking at everybody." Bonding, then, depends in part on the animals' actions, and on human responses and expectations (Arluke, 1988; Ginsburg and Hiestand, 1992).

Technicians spend more time each day with their animals, so are perhaps more likely to develop bonds. Because of those friendships, many technicians felt, they were the best ones to handle the animals or even to cull them—"it's better if I do it, because the animals see me as a friend." Thus, one technician explained, "The ferrets knew when they were going to be operated on or to be put down, and no matter how hard you tried to hide it they knew. You also realized that you were the best person to do it because they trusted you and that you knew how to handle them." Here again, the attitude of technicians reflects their ambivalence about objectifying the animals in their charge: they may be part of the process of science but they are also "friends."

A final difference has to do with work and identity. Research scientists tend to talk about their work and career progression in terms of solving biomedical problems. They spoke in general terms, such as "doing electrophysiological studies attempting to relate brain activity to pituitary function" or, if they did specifically refer to using animals, it was because they were trying to "find an animal model for specific human genetic disease." As we have noted, research scientists have learned, early on in their training, to establish distance and to see animals in terms of data; many, moreover, have only limited contact with the animals in the animal house, interacting with an animal only when it is anesthetized and ready for surgery. For them, their identity as scientist is more focused on the problems to be solved.

Technicians too sometimes construct their identities through the scientific problems to be solved, at least in a general sense; that is, they might identify themselves as part of a wider endeavor of scientific research. But their identity is also based on their relationship with the animals, emphasizing that a "love of animals" is what started them in the job. It was this love of animals which, some felt, enabled them to do the job well—to handle ferrets, for instance. Some spoke of the work with animals being a prime reason for applying for the job in the first place, and then being pleasantly surprised when they were shown around the animal unit—they were quite happy with the routine care of stock animals, even if they felt uneasy about some of the experimental procedures.

Coping Strategies and Emerging Identities

Research scientists and animal technicians are both engaged in the same overall enterprise—the pursuit of scientific research. There are several ways in which that common goal influences attitudes toward work and animals, but there are differences too, which reflect and feed back into relationships between people and between people and animals. As we have seen, for example, technicians particularly see themselves as aligned with the animals, or even portray them as their friends.

Many researchers, however, are not spending all their time near the animals, and are much less likely to talk about bonding with them. The physical separation of the animal house from the laboratories means that much of their work will be done elsewhere. Experimental design, too, restricts the extent of human interaction with an animal. In, for example, "acute" studies, the animal will be anesthetized the whole time it is in the lab, and will then be euthanized. Researchers can spend little or no time with the conscious animal in such cases. Not so the technicians, however, whose job entails caring for all the animals, including those destined for acute studies, and who will therefore never come back. For technicians, and for those researchers who do most of their work in the animal house, individualizing certain animals and making them into friends is more likely.

What these points of similarity and dissimilarity underline, however, is that doing animal experiments requires moral justification and psychological adjustment. How people use different coping strategies for dealing with such moral difficulties depends upon their position in the social structure of science. Technicians often found it harder to treat animals as objects because they commonly lacked prior research experience, and had more direct contact with animals. So their coping strategies included seeing themselves as buffers and treating the animals as friends. Research scientists new to the job (doctoral students, say) sometimes referred to their own disquiet, just as students did when confronting dissection for the first time. Experienced scientists and administrators were much less likely to debate these issues, more willing to portray the research in terms of potential medical benefits. But this too can be seen as a coping strategy, even if by that stage of career it is a well-rehearsed one.

Part of the means by which researchers learn to cope with the dilemmas is a form of moral derogation; that is, the animals become cast as objects or data. The everyday practice of science is clearly different from the ideal image of it as dispassionate and unemotional. It is not surprising, then, that the key theme running through our interviews with scientists and technicians alike is ambivalence: however well they learn the psychological mechanisms of distancing and treating the lab animal as just another tool of the trade, there is for many lab workers a naturalistic animal—a furry friend—there too.

Scientists' distancing serves a purpose for their beliefs and their own sense of self. Distancing does not always work, however, and scientists are increasingly acknowledging that there is sometimes a clear bond that develops between experimental animal and human (see chapters in Davis and Balfour, 1992). That bond has implications, of course, not only for animal welfare but also for science; how the human and the animal interact can often have effects on the behavior or physiology of either participant. In that sense, the animal and the human can both be said to shape the science.

In our interviews, many respondents referred to relationships with animals in their lives outside of the lab, particularly with pets in the home—indeed, this is a point often stressed by spokespeople for research advocacy groups (see chapter 8), who may seek to publicize their pet-keeping to imply that they are just "ordinary folk." Relationships with pets are likely to be with what is seen as a naturalistic animal, or a nonhuman member of the family—a quite different relationship from that with animals in the lab.

Studies of psychological profiles and attitudes of pet owners have indicated, perhaps not surprisingly, that pet owners are more likely to rate using animals in research as unacceptable than those who do not own pets (Driscoll, 1992; Hagelin et al., 2002). Equally unsurprising is the finding that those people who have a strong belief in animal mind are less likely to approve of using animals (Knight et al., 2004). Historically, too, the growth of pet ownership throughout the early modern period was a factor which contributed substantially to greater sensitivity toward animal suffering (Thomas, 1983). Clearly, many of the people who have such beliefs about animal capabilities are unlikely to want to work in animal experimentation. But some do, and they have to find a way through the morass of conflicting feelings and beliefs.

As seen in this and the previous chapter, we learn many conflicting ideas about animals and, in turn, have emotional responses to these conflicts, which vary in different cultures. In Western industrialized countries, images of animals as cuddly friends or as cute quasi-human cartoon figures contrast in children's early experiences with what they sooner or later learn about—say, animals becoming meat. But there is a particular challenge from the kind of objectification that inheres in science education and practice. The implications for human identities are that a person must learn to differentiate between one sense of self, perhaps in relation to his/her own companion animals, and another sense of self, as a scientist pursuing objectivity.

If identities emerge from our engagement in particular networks, then becoming a person who uses animals in experiments while simultaneously living with them at home must often mean shifting networks. On the one hand, there are those networks in which animals figure as subjects, and which structure many of our experiences with animals in the home, as walkers or breeders of dogs or whatever;[6] on the other hand, scientific practices structure human-animal and human-human relationships in quite different ways. In the former, there is an acknowledged intersubjectivity, permitting particular ways of responding to/with animals and ascribing to them particular traits, which in turn are reflected in associated human networks (shared assumptions made, for instance, by members of a breeding society or a dog training class). But in the case of scientific practice, relationships with the animal have traditionally been downplayed, or the animal's subjectivity denied.[7] Learning to do science thus means learning to separate these two ways of relating to animals, becoming socialized into the accepted networks of scientific communities. This is never easy.

VI

Organizing and Regulating Lab Work

[The U.K.] law safeguards laboratory animal welfare while allowing important medical research to continue. These controls are widely regarded as the strictest in the world.

—Research Defence Society, webpage

[The claim] that the UK has the strictest laws in the world to protect laboratory animals . . . is an empty claim. What is indisputable is that the main function of the Act . . . is not to protect laboratory animals, but to protect animal researchers by allowing them to subject lab animals to the sort of treatment which would be illegal outside the laboratory.

—British Union for the Abolition of Vivisection, webpage

Whatever feelings lab workers have about animals they use or care for, they must situate these within a wider context—in particular, people with whom they work, and the institutions (such as the law) that frame their research. The law in most developed countries exerts some control over what scientists can or cannot do, although its impact is open to very different interpretations, as the quotations above indicate. Not surprisingly, all the scientists we interviewed, both American and British, were broadly in agreement that their national laws regulated science and helped to protect animal welfare; British scientists generally agreed that theirs was one of the strictest laws.

Learning to do science is not only a question of acquiring knowledge at the lab bench; it is also about learning to operate in a complex set of regulatory and social networks that support scientific practices. In this chapter, we examine some of these, beginning with social networks in lab work and how people perceive them, before considering the legal framework within which these human relationships operate. The law is what, in principle, opens up possibilities for control, even if anti-vivisectionist organizations see it as a scientists' charter.[1]

Apart from their dealings and feelings about animals, lab workers must also negotiate social boundaries between themselves. The most obvious of these is the division of labor between research scientists and animal technicians; as we noted in the previous chapter, these two groups inevitably have different experiences with laboratory animals and different relationships to the research. The social division of labor is part of the context of scientific practice, deeply tied to how research is structured in space and time. That in turn has impact upon how lab workers experience their work, and whom they consider to be part of their occupational group, with implications for how they understand external hostility toward animal experiments.

Doing science, like other forms of work, is highly organized in time and space. We noted in chapter 2 how labs became more spatially organized as scientific work became more industrialized through the twentieth century. That process of industrialization meant that science became increasingly capital-intensive, which in turn facilitated a greater division of labor and specialization between scientists and between scientists and other lab workers. Describing this division of labor, Albury and Schwarz (1982) noted several consequences; for example, scientists in "basic" research cannot always see how their work might be used. This in turn means that they must work harder to defend their research to outsiders (if using animals, that often means extrapolating to potential human benefit, even if such benefit might be a long way off); the pronounced division of labor between scientists also means that "basic" scientists rarely talk to those working in applied fields. As we noted in chapter 2, one consequence of that is that there is remarkably little crossover—despite the rhetorical claims about potential benefit—between animal-based experiments and the clinical studies that are purportedly linked to them.

Another consequence of greater division of labor is that a system of contract labor has developed, particularly over the last fifty years, with many young researchers now on short-term contracts for several years. Thus, although it will be the principal investigator who oversees a project, it will in practice be the pool of contract labor—the graduate and postgraduate students—who have the closest contact with technicians in the animal house. The work of junior scientists, suggested Albury and Schwarz, broadly resembles a production line—even to the extent that the organism under study can itself be divided up between different groups of workers. Thus, Pickvance (1976) described how the nematode he worked on as a doctoral student was divided—he got the eggs—such that only the supervisor had any sense of the whole animal. That emphasis has not changed, and is arguably sharpened by the growth of genome mapping, which further subdivides how we understand animal bodies.

Dickens (1996) notes that animals have long been part of the division of labor, but that role has intensified. Using the example of the stockmen and women who have traditionally cared for their agricultural animals, he argues that this lifelong management has begun to give way to a more fragmented management system, and a number of specialized sciences. In such situations, animals become increasingly like commodities, and the potential for them to relate to specific humans is lessened; we saw in earlier chapters how the standardization of laboratory animals and separation of some behind barriers have contributed to this process. It is the technicians' job to care for the animals; but, seen from another perspective, it is also their job to manage this fragmentation of the lives of their charges.

As science became more industrialized, animal houses became more common as separate units, and the work involved in animal experiments became more subdivided. Once laboratories and animal houses separated, so too did the technical work associated with them. The specific category of animal technicians, with their own professional organizations, arose in the second half of the twentieth century, as animal production became more routinized and allocated to specific locations.[2] So technicians are separated from research scientists not only by profession and training (the hierarchical division of labor) but also by where they principally work.

That spatial separation is further facilitated by modern technologies, permitting communication by fax or email. This means that labs and animal units could be some distance apart, with the result that research scientists themselves seldom need enter the animal holding unit—or even see one of the animals whose tissues they use. For example, one laboratory in Britain where we conducted interviews was several miles away from its supplying animal house; the research focused on studies of cancer, and scientists sent the research protocol by fax to the technicians. The technicians were then required to monitor the progression of tumors in the mice and send tissue samples to the laboratory scientists by courier. In this case, there is a very marked separation of research scientists and technicians, as well as of scientists and animals. One researcher there described colleagues who "have been out [to the animal house] four times in four years and probably used more mice than the rest of the lab put together. It is partly up to the individual how much you go out there. Because they never actually see the mice." Others emphasized how they might simply talk to researchers on the phone and then pass animal cages through doors, with very little direct contact.

Making a similar point, another scientist in this unit pointed to the centrality of the written report in the lab's production:

> . . . if you are looking at tumor growth . . . the animal technician up at the other unit will go through it and write a [postmortem] report, and they will say well it has spread to the liver or spread to the lungs. And you will get this report back and say well the group that had these cells has spread more than the group that had those cells. So you won't actually see a mouse, ever.

Although these technicians arguably had a more interesting and responsible job than some, doing the postmortem examinations and compiling data, they were also disgruntled that several scientists simply never saw the animals. This, they felt, was morally wrong.

Thus, the marked physical separation of animals and scientists also means a separation of people: if some scientists never see the animals, then they will rarely engage in face-to-face encounters with colleagues working with the animals on a daily basis. This particular institution was an extreme case, admittedly; nevertheless, the very separation of animal houses from the rest of the institutions which they serve—however necessary in terms of cost, health and safety, and protection from attack—reinforces the division of labor among people and the separation of human from animal. While several scientists and technicians were at pains to stress the cordiality of human relationships at their institution, those relationships were structured through occupational hierarchies as well as the physical spaces constraining them. According to technicians, their relationships with scientists were generally good, but occasionally "stroppy" or difficult, especially when technicians felt that their views were not being taken seriously.

Demarcating Social Boundaries

As in other professions, social boundaries are clearly maintained in labs and animal houses. Interviewees told us how particular jobs were demarcated. As we saw in previous chapters, technicians, for example, sometimes saw their role as a buffer between themselves and the animals, occasionally intervening and telling the scientists that an animal should be humanely killed or suggesting alterations to the experimental design. But on the whole, they saw themselves as having a very specific role and location:

> . . . the aftercare is our specialist work . . . things are opening up and
> . . . they want technicians to do more and more involvement in the
> research. Now I don't agree with that, that's my own personal opinion.
> I think animal technicians are animal technicians, not animal research
> technicians, and that has always been [like that]. But you see, I see our
> field in the animal house. The animal house is where we work and the
> animals under that animal house are in our care.

For this technician, there was a very clear sense that what animal technicians did was clearly defined as a separate job from mainstream research. And not only did she see it as a definite service role, but it was also a job that had its specific location: the animal house.

Several technicians were at pains to emphasize the significance of their role. Scientists, too, sometimes alluded to different areas of expertise. One British scientist, for example, working in cancer research, thought that having separate spheres was good, since "Ninety-nine per cent of scientists don't know anything about whether an animal is feeling pain or not. These guys [technicians] are trained specifically for that purpose. I don't think I have ever come across anybody who would have that skill as a scientist to know when to stop an experiment, or when an experiment should be terminated and to say, right, that animal now has to be killed. . . . And no scientist knows that—it is the animal technicians who know that and that is mainly from experience." Technicians often agreed that they were the best ones to judge when an animal was in so much pain it should be put down.

Many technicians saw themselves as less separated from the research, however. When asked about how they saw themselves in relation to the public controversy, for example, many referred to how laypeople "don't know what goes in science," thus aligning themselves with the scientists as insiders to the research enterprise. Like scientists, they spoke of how they could not always divulge publicly what they did for a living (a theme we expand on in chapter 8).

Technicians are also insiders in the more specific sense that they are inside the *particular* space that is the animal house (or the lab if they are lab technicians). Unlike scientists, their career structure is less tied up with moving on to other institutions in pursuit of specific career goals. Writing about how knowledge is created in laboratories, Latour and Woolgar (1979) point out that the more successful scientists become, the less they are involved with a particular lab. They are more likely to move on to further their careers. In other words, they themselves become—like the lab—epitomes of placelessness. By contrast, animal technicians and their work are more bounded and connected to *specific* places and practices.

Yet technicians are also outsiders, in that they do not necessarily participate in the decision-making and laboratory practices that guide experimental projects. Instead, they "wait for the animals to come back" from surgery and care for them afterward. So in some ways the technicians, while acknowledging their role in the research process, are also closer to the outside world, in the sense that they are not directly party to the cycles of knowledge production and ratification. Indeed, they sometimes articulated views locating themselves as outsiders—emphasizing their own ignorance of what went on in particular experiments, for instance, or sometimes aligning themselves with overtly "animal rights" positions in our interviews.

So, while several technicians stressed the importance of their role in the research process, others emphasized that they did not know, or want to know, what the purpose of the experiments was—or alluded to ways in which the scientists did not tell them. One technician never asked why animals were being used, saying that she "never thought about asking to be honest . . . we sort of do as we are told." This kind of willful not knowing is an example of what Michael (1992) called a "discourse of ignorance" rooted in the division of labor ("it's not my job"). Coupled with technicians' desire to demarcate their own areas of expertise—caring for animals—this creates a defense which partially avoids the center of the ethical controversy. By choosing not to know what is happening or why, while simultaneously stressing the importance of their role in "keeping the animals happy," these technicians were clearly defining themselves as outsiders to what they sometimes portrayed as the ethically dubious practices of animal experimentation.

Technicians, however, also recognized that they too were part of the scientific process that comes in for public criticism. This dilemma was often dealt with by emphasizing how animals were well cared-for in *their* institution, while abuses went on "elsewhere" by "others"—we saw an instance of that in the last chapter in relation to research done in other countries. In other words, they were saying that some things done to animals in labs were indeed morally dubious, but that those practices did not happen in the part of science they inhabited.

Scientists and technicians concurred in this boundary-making. Both saw themselves as inside science and inside particular moral boundaries; it was outsiders (outside science or at other institutions) who should be morally derogated. This is, in effect, a way of acknowledging that there is a moral problem about using animals but implying that they, in that particular place, were doing their best. This rhetorical move does two things: first, it creates what we have called a socioethical domain, outside of which are people who do not "do things right" by animals. Second, it provides a powerful connection between groups of people—animal technicians and high-ranking scientists—who otherwise occupy very different places in occupational hierarchies: they are bonded by shared understanding of what counts as "good" animal care. This does not necessarily mean that technicians always defended practice in their own institutions (although senior scientists usually did). Some certainly admitted that they had seen things that they thought were not good practice or were downright cruel. But on the whole, most were keen to let us know that "things are always done well here" (see Michael and Birke, 1994a).

Things were not, however, always done properly elsewhere, they informed us. Apart from the "foreigners" referred to in the previous chapter, it can, for university scientists, also mean industrial laboratories conducting toxicity tests (especially on cosmetics); or it may mean a different area of biomedical research or personnel (clinically trained surgeons conducting surgical experiments on dogs, for instance). When these "others" and their practices were invoked, it usually entailed reference to "bad" science, contrasted to the "good" science carried out in the speaker's own lab. Only good science, we should infer, can justify the use of animals.

As we saw in previous chapters, people working in science find it hard to come to terms with using animals. Once working in labs, they continue to find ways to deal with the moral dilemma; part of doing so is to deploy rhetorical strategies that defend their personal choices and locate them in specific communities of researchers. Outsiders to such communities are "other," often cast as irrational and unable to appreciate the subtleties of science. Animal technicians occupy an interesting dual position: in relation to the wider public and anti-vivisectionist arguments, they are clearly insiders, and are seen as such by the research scientists. Yet technicians themselves are more ambivalent: in relation to the public, they portray themselves as insiders who are a necessary part of the research process. But they also see themselves as partially aligned *against* the scientists, as buffers between the scientists and the animals they use.

The rhetorical strategies employed by research scientists, then, both include and exclude technicians, thus helping to maintain social boundaries. Both, however, categorize opponents as irrational and invoke the powerful metaphor of being shut inside a fortress—those outside it are mired in the forces of irrationality—which takes the literal form of multiply locked doors to animal houses to keep people out. In that sense, then, the psychological and rhetorical process of demarcating "insiderness" serves not only to support social boundaries but physical ones too.

Industrialization and fragmentation of work have also promoted an increased reliance on the law to arbitrate ethics. When the first British legislation was passed, in the 1870s, scientists could argue in print how they were the best judges of what happened to animals, and that they would not like anyone telling them what to do (e.g., through legislative means).[3] But science has, over the last century, become much more regulated through modern regulation and government. Describing these changes, Ewick and Silbey (2003) note the transition from earlier ways of doing science:

> Rather than the scientist inviting acquaintances to his home and relying on conventional morality to secure trustworthy witnesses to scientific experiments, the contemporary research laboratory is a space governed by a network of laws, regulations, and rules helping to produce a specific kind of subject: a particular kind of scientists and a particular kind of science. In the spatial regulation of science, processes of social control are largely internalized, sustaining science and the scientists' authority and capacity for autonomy and self-governance. (p. 2 of website paper)

Ewick and Silbey's own concern was with how the law operates in relation to science, focusing on health and safety legislation in U.S. labs. The first response of scientists interviewed, they comment, is that "there will be little to talk about"—in other words, these scientists believed that the law was irrelevant to the activities of science. They acknowledged that there were regulations controlling what they could or could not do (with animals, for example), but these constraints were dismissed as being external to the process of discovery itself. Whatever scientists believe about their work, however, the law profoundly shapes what they do. In the course of the interviews, they emphasized various ways in which legal constraints limit and constrain their movements in laboratory spaces—through requirements to contain hazards, for example, which demand barriers. What these requirements have done in turn is to prevent scientists from "living their lives" in the lab; now, separate areas for eating are demanded by law. One scientist rued this change, telling the interviewers that he used to like "the life; I was a lab rat. I enjoyed being in the lab" (Ewick and Silbey, 2003). Now, however, because of legal mandates and controls, that way of life is no longer possible and has become fragmented.

While the separation of animal houses from laboratories began long before legal controls were instituted in many countries, the law certainly facilitates the separation. It also supports it in that health and safety legislation specifies conditions for the containment of potential disease transmission between humans and animals. So scientists working with animals are directly constrained because they must abide by the legislation and its bureaucratic demands, and they are constrained indirectly because the law acts to shape the spaces in which science is done. In that sense, we can say that the social relationships between lab workers, between lab workers and animals, and between lab workers and those outside the lab are all shaped by the cultural frameworks within which the science takes place.

The Wider Framework: Regulating Research

Whatever forces help to structure science from the inside, it is also fundamentally shaped by wider society, particularly the legal and political framework. There are many laws which impact upon scientific research indirectly—legislation covering health and safety at work, toxicological testing, wildlife, employment practice, and so on. But many countries also have specific legislation governing the care and use of animals in scientific research that controls how animals can be used and sets limits on their potential pain and suffering. Apart from legislative requirements, researchers must usually consider institutional demands—from the organizations within which they work and from professional bodies to which they belong. Professional organizations, such as learned societies, might for example require that all papers published in their specific journal indicate how ethical guidelines were followed; papers that have not done so or which pose ethical questions are then unlikely to be published in that journal.[4] Grant-giving agencies, too, usually demand evidence that proposed research will adhere to ethical guidelines and/or legislation.

In relation to animal laboratories, the law matters a great deal, putting constraints on what can be done. The often stringent requirements of statutes can stipulate anything from animal-stocking densities to the frequency of air changes within an animal room. The very fact that there is legal control over animal care and use also means that animal houses must be clearly demarcated from other laboratory spaces, often with high-security locks in between, and their internal spaces separated into specific functions (separating, for example, breeding and operational rooms). According to British law, for example, individual premises must be licensed, with separate licensing for breeding/supplying institutions and those carrying out scientific procedures.

Different countries inevitably have quite different legislative frameworks, depending on their own constitutional status, political culture, and membership of larger structures (such as the countries within the European Union).[5] Broadly, however, there are two types, which have been characterized as "at the top" or "at the bottom" (Bradshaw, 2002). The U.K., for example, employs a "top-down" strategy, with a statute enforced and controlled by a publicly accountable government authority, the Home Office, while Australia by contrast uses an "at the bottom" strategy, with controls enforced by the state being implemented mainly through institutions themselves.

Both the United States and Britain have enacted legislation which governs in some ways the use of animals in scientific research, and both are likely to require ethical justifications (for both animals and humans) for grant-giving agencies. In both countries, too, the legislation basically reflects a compromise position that does not make experimentation illegal, but reflects some of the opponents' concerns. For example, the precursor to current legislation in both countries gave dogs and cats special consideration (Groves, 1994), reflecting public anxieties about the potential fate of pets that found themselves in animal shelters.

Britain was the first country to enact specific legislation covering animal use in 1876. The Cruelty to Animals Act required a system of licensing of individuals and institutions, but was eventually superseded by the 1986 Animals (Scientific Procedures) Act.[6] Both acts have operated through the government Home Office, with inspectors to oversee the working of the law. The current act specifies a tripartite licensing system, such that not only the institution and the researcher must be licensed, but also the specific project. It also demands high levels of control over pain or suffering[7] and specifies conditions of animal husbandry. More recently (from 1999), there has been a requirement that institutions put into effect an ethical review process.

The ways in which the legislation works differs, however. In the U.S., the Animal Welfare Act of 1966 (updated several times since) sets standards for animal care and husbandry, but specifically does not apply to the performance of experiments using animals (unlike the British legislation).[8] Rather, such controls operate through the requirements and funding controls of the grant-giving agencies. Thus, research funded by the National Institutes of Health is also regulated through the U.S. Public Health Service Guide for the Care and Use of Laboratory Animals and the Health Research Extension Act of 1985. These specify that researchers use due care toward animals and establish acceptable protocols; the primary system of regulation in the U.S., however, is at the institutional level, through the operation of institutional ethics committees (institutional animal care and use committees, or IACUCs). In this sense, it is a system more akin to the Australian one, with control "from the bottom."[9]

Both sets of regulation generally specify vertebrates. But the U.S. federal

Animal Welfare Act has famously excluded rats and mice from its purview (an exclusion which has repeatedly been challenged in the courts), while U.K. legislators have added cephalopod mollusks (squid and octopus) to the vertebrates covered by the 1986 Animals (Scientific Procedures) Act.

Both countries specify that a veterinarian must be involved in the control process. In the United States, this works through the U.S. Department of Agriculture (USDA) requirement that registered institutions must have a minimum of three members, one of whom is a veterinarian, and one of whom is a person not affiliated with the institution. The British system similarly specifies a "named veterinarian" for each institution.

Whether operated through top-down control or via institutions, the systems of control in both countries are intended to provide proper facilities, including adequate environments and housing for animals (definitions of which are constantly under review),[10] as well as veterinary care and training for those handling animals. What does differ is the point of emphasis: in the U.S., while the IACUCs do have some say over experimental protocols, the emphasis is more on animal maintenance. In the U.K., there is more emphasis on experimental procedures, which have to be negotiated in detail with Home Office inspectors.

On the whole, the various people involved in policing legislation in both countries broadly agree with what it tries to do. In the British interviews, for instance, scientists told us many times of how well they got on with the government-appointed inspectors, or of how they thought that the change in the law (in the mid-1980s) was a good thing. But enacting policy is also problematic. One area of disagreement in both locations was the way that the law and government guidelines made demands of animal husbandry that some commentators felt were inappropriate. For instance, a British researcher who specialized in animal behavior felt that maintaining a constant temperature in mouse rooms was not necessarily appropriate; mice like to huddle and make nests out of soft material, he pointed out, thus keeping their own temperatures under control in ways more biologically relevant. And he wondered whether constant environmental conditions were even desirable for the animals' physiology.

Groves (1994) reports similar arguments in American labs. IACUCs may demand high standards of cleanliness that researchers feel is often quite inappropriate; one researcher, an expert in echolocation in bats, continually resisted the IACUC demands to clean up his bat facility because every time he did, he found several animals died; living in mucky caves "is the way bats live," he argued. Space is another point of contention: whatever federal legislation or institutional committees may say about minimum space, veterinarians or scientists sometimes feel those demands are pointless if all the animals huddle in one corner. So what these scientists are saying is that, yes, the law is beneficial, but the way that it operates is not optimal; in so doing, they posit their own expertise, as scientists with knowledge, say, of animal behavior, against that enshrined in the law.

Whatever the details, the law is widely seen as desirable because many researchers believe it controls excessive cruelty. Paradoxically, several interviewees also portrayed the change in legislation as having made little difference, because things were "done well" before, as one scientist commented: "When I say we haven't changed our practice, I jolly well know we were doing things correctly before the new Act came in. But I think it has provided a degree of recognition that we are doing what we ought to be doing." Another said that he had been "a bit concerned that things would be so tightly worded in the project license that it would eliminate that creative potential . . . but I think we've managed to come to terms with these things adequately."

Yet the law is also perceived as a nuisance, in that it involves considerable amounts of paperwork. "The worst thing about the . . . legislation," commented one British researcher, "is that I've spent hours trying to understand the information and fill in the forms. Even when I've got to the point where I think I have, I then have to ring up the Inspector to make sure that I understand right."[11]

Another felt that a great deal changed after the 1986 act came into force, but at the same time,

> The way in which we handle and use the animal has not changed at all. All that's changed is the volume of paperwork and the administration that's involved. . . . That means that although in some ways it's good because it means that you have got a number for an animal and a handle on that particular animal . . . rather than it being just one of a batch, it's terribly time consuming to keep the records, especially for the transgenic work . . . it takes forever.

This bureaucracy, in turn, was seen to hamper research without necessarily improving animal welfare, as another British scientist argued:

> The actual work I do hasn't changed at all but then we were fully licensed before. . . . The difference it does make is that I'm now scared to death of doing anything because we've got inspectors coming round and if there is what I consider to be a trivial mistake on a label of a box then the inspector will be on you like a ton of bricks. . . . It's certainly given me a lot more paperwork to do now. . . . I don't think it's improved animal welfare one iota. . . . I think it's a bureaucratic exercise.

Irritation with the paperwork is a feature of U.S. science, too: when changes were introduced to the Animal Welfare Act, one scientist commented that enacting legislation makes it look like scientists were "forced to take measures we were doing already."[12] An American veterinarian similarly felt the law and the paperwork were a burden, and that researchers thought his role was to help them find a way around legislative requirements: "I'm trying to help them," he commented. "But they don't look at it that way. They look at me as being a barricade they have to either run through, climb over, walk around or whatever—that I'm not that helpful, I'm part of the problem" (Groves, 1994).

Putting up barriers can take the form of overt resistance: Groves further cites an IACUC member who was getting very frustrated with a scientist who persistently played games by not giving any information about his research, so forcing the committee member to track him down and complaining each time about how "stupid" the regulations were. And not only do some researchers complain about and resist the authority of the IACUC, so too does the IACUC itself sometimes resist USDA inspectors, complaining that the inspectors have "a lot of book learning and not a lot of practical experience" (Groves, 1994).

Yet while scientists complain often about the hassles of form-filling, they also consistently told us that the regulations were, on the whole, a good thing, not least because they forced scientists to think through the issues. In our interviews, we expected that older scientists who did animal experiments before and after the 1986 act in Britain would see it as some kind of threshold. Instead, what we found was that they saw it as both welcome, leading to "increased awareness," and yet simultaneously making no difference. It seemed important to these respondents to convey a belief that British animal experimentation was part of a tradition that has directed much effort and many resources to caring for animals.

One scientist summarized this tension; he believed that "we always were sensitive with regard to the treatment of the animals. . . . I think the law has made people now sit down and think . . . go and get involved in tissue culture." Several others pointed out that they had "always done" spot inspections in their labs, insisting that the overall climate of concern for standards of animal care had changed little.

Yet many also emphasized the positive impact of legislative change that makes people "think harder." If researchers working with animals are forced to reflect more, they will "increase awareness," which will lead to better science in the end, they argued: "By having to put together a very large project application, is to focus your thoughts. I think that's extremely useful. It's a pain in the arse as well." Another respondent noted: "There has certainly been more time spent, and more thought about what we are doing."

This "thinking harder" is, however, not only a product of legislative requirements: it also reflects growing consensus in the scientific community that the issues must be discussed more widely and the welfare of the animals given higher priority. Thus, in the United States, there were changes in the way that IACUCs fulfilled their mandate. Partly this was a result of developments in science, such as escalating use of transgenic animals and the development of in vitro alternatives to the production of monoclonal antibodies. But partly it also reflected growing concern during the 1990s to improve the animal handling skills of scientists, and to provide environmental enrichment, especially for primates (ARENA/OLAW, 2002).

Although the primary route of control over experimentation in the U.K. is statutory, rather than local, an additional system of local committees was set up in the late 1990s to provide ethical review at the institutional level. The ethical review process (ERP) began in 1999 with the intention of promoting positive attitudes to animal welfare and due consideration of ethical questions involved in animal experiments. This provided another change in how the systems of control operate in Britain. In general, most people involved with animal experiments believe the ERP has been beneficial, or potentially so; such local committees can, for example, facilitate proposals by providing a forum at the institutional level and can provide expertise on specific local conditions. They can also, importantly, provide a voice for people caring for animals such as technicians, who will have particular expertise and awareness of local conditions (Boyd Group, 2001).

However, several members of the Boyd Group (a forum for discussing issues related to animal experiments) also felt that the ERP added to the burden of paperwork and, in the case of researchers who are also clinicians, took time away from patient care. Others pointed out that having another committee could potentially slow down applications for project licenses, thereby slowing further the time between planning and doing the research.

Similar responses were made to a detailed survey of scientists and others involved in animal experiments after the ERP came into force (Purchase and Nedeva, 2002). Although most senior people surveyed understood the ERP and what it was meant to do, fewer junior scientists did: only 55% of junior researchers[13] knew there were written procedures, for example, while 80% of senior scientists and associated veterinarians knew so. As Purchase and Nedeva point out, the problem here is not whether institutions do have the required written procedures, but whether researchers in those establishments know that they do.

More than two-thirds of those sampled felt that their work was monitored "to make sure it meets the necessary standards." But the corollary is that about 20%[14] of licensees said that the statement was false—in other words, that their work was not monitored. Similarly, while most felt that the training they received in animal welfare was good or excellent, training in ethics was generally felt to be of lower quality (Purchase and Nedeva, 2004). Yet, despite a few who thought there were some unethical procedures not reported to the ERP committee, respondents were almost unanimous in believing that animal care in their organization was excellent, good, or adequate (99% of respondents answered in these categories).

Like the scientists we interviewed, many of the respondents in the post-ERP survey generally felt that things were the same before the ERP was introduced—"we were doing it all right before the change." But some of the survey respondents also felt that ethical quality had been substantially improved by the ERP. So just like the changes wrought when the new law came into force in 1986, the introduction of another level of control in 1999 was seen as "change and no change."

What the more recent survey underlines, however, is the differences between senior and junior staff. Not only were junior staff less likely to be aware of procedures, they were less likely to believe that the ERP and those involved with its operation actively supported the use of alternatives. Although the stated purpose of the ERP is to promote welfare and the use of the "three R's"—reduction, refinement, and replacement—it was primarily the senior staff who understood this (over two-thirds of those sampled). While most of the personal license holders sampled (usually more junior staff) said that they were interested in alternatives, only a third believed that "Senior people are actively involved in promoting the use of alternatives" (Purchase and Nedeva, 2001).

The law, in conjunction with changing public attitudes towards animals, shapes how scientists use animals, and what they believe about their work. What is very clear from our interviews and other studies, such as the survey of responses to ethical review, is that scientists are now exposed to a climate in which ethics and animal welfare are increasingly under discussion. Most see that as a good thing, exposing the work to critical evaluation, and thus enabling them to "think harder" about the issues.

That climate, however, is double-edged. On the one hand, scientists say that it makes them more focused, more aware of issues. But in saying that, they necessarily imply that they were *less* aware previously, thus laying themselves open to potential criticism. The way out of that dilemma is to claim both that things have improved *and* that they were done well before. That was a theme in our interviews, particularly in Britain, where the change in law in 1986 provided a definitive change; it was a theme echoed in subsequent surveys.

This "climate of discussion" takes place against a background of social hierarchy, however. Junior scientists and technicians are less able to speak openly and may feel obliged to take the view of others in the institution or lab. Groves (1994), in his study of laboratory animal policies in the U.S., points out how the hierarchical nature of the review process makes it more likely that infringements would go unreported—which is the opposite of what the law is intended to do (also see Purchase and Nedeva, 2002). Despite all the "focusing of thoughts" brought about by this new climate of discussion, some lab workers still feel that it is not possible to report infringements to outside authorities.

Laws directly relating to laboratory practices and laboratory animal care themselves take place in wider contexts. There are, for example, a wide range of other legal requirements that impose constraints on what can or cannot be done in laboratories. But because laboratory animals are the means and not the end of many other legal requirements, there are often no safeguards for animal welfare. Purchase (1999) discusses, for example, safety assessment of new chemicals. Toxicity testing is essential to protect human health and usually relies on the results of animal experiments. Yet national and international proposals for toxicity testing are not subject to any form of ethical review.

The example Purchase analyzes is the international assessment of chemicals such as potential endocrine disrupters. Proposals for assessment of these from the U.S. Environmental Protection Agency are, he points out, "extravagant in the use of animals" (ibid., p. 141), requiring between 0.6 and 1.2 million animals for each 1,000 chemicals tested. As he notes, the chemical and pharmaceutical industries are international, so that testing requirements of one national agency such as the EPA has a considerable impact in other countries.

Funding agencies are another constraint shaping what happens in animal experiments. Major funding agencies in the U.S. and the U.K. stipulate that while they recognize that animals must be used for some kinds of research, there is a need to reduce their numbers and to support ethical guidelines (although in neither country is there systematic means of auditing funded proposals after the event). In 2004, for example, a National Centre for Replacement, Refinement and Reduction of Animals in Research (NC3Rs) was set up in Britain, supported by leading agencies such as the Medical Research

Council and the Biotechnology and Biological Sciences Research Council. Grant-giving agencies are thus sources of regulatory control, but they also see themselves as contributing to a growing "culture of care."[15]

The need to develop a culture of care reflects mounting concerns about animal abilities and suffering, and the growth of understanding within science about animal welfare. Increasingly, funds are being made available for research into animal welfare within laboratory spaces, with the aim of improving the lives of animal inhabitants—better cage designs more suited to animal needs, for instance, or new ways of providing enrichment. Although this is inevitably criticized by many animal rights activists (it does nothing to end the practice), many others see it as a welcome move toward making the lives of laboratory animals less stressful. All this in turn takes place in a changing climate of public opinion about animals.

Yet if there is now more public awareness of animal consciousness than there was, say, fifty years ago, there is also greater hostility toward science. That is the double bind: as public agencies such as grant-giving bodies increasingly acknowledge the need to create a culture of care and insist on ethical review, they draw attention to the fact that animal experiments *need* ethical debate. The scientific community has had to respond to the arguments of animal rights protagonists in order to "win the hearts and minds" of the public. As a result, there is now considerably more public discussion of the issues, and more public agencies openly acknowledge them. As we will see in the next section, however, this remains problematic: more public knowledge does not necessarily mean greater acceptance.

Managing Identities

Throughout this part of the book we have looked at various ways in which scientists and other lab workers learn to deal with the moral dilemmas of using animals in experiments. As they become more socialized into the networks that comprise science, they learn to negotiate these dilemmas. These processes are in turn crucial for people's identities as researchers, technicians, and so on. But learning to work within networks and taking on specific identities does not remove the ambivalence, and we have also seen how lab workers find various ways of expressing that ambivalence.

People working in animal research operate in multiple networks, however; a scientist can learn usual practices (such as dissection) and take for granted various underlying assumptions, yet must simultaneously position him- or herself in relation to other networks, such as the wider public and regulatory authorities. Each of these makes different demands on the person's identity and ways of presenting him- or herself. Thus, scientists straddle many networks, and they must somehow "manage" the resulting contradictions and ambiguities. "Identity" under these circumstances is not only about identification within particular networks, but the managed movement between and across networks. In the next part, we will turn to how scientists achieve these transitions, but particularly how they manage the apparent chasm between their own understandings within the communities of science and the perceptions of the wider public.

PART III

CONFRONTING THE PUBLIC

When asked to speak about their research, researchers necessarily address a range of audiences. In conducting our interviews, we constituted one audience, a primarily academic one. Audiences might also more narrowly comprise other scientists through specialized publications, anti-vivisection organizations, or a wider, scientifically literate section of the public (addressed, for example, through articles in magazines like *New Scientist*). Sometimes, the audience addressed is "the public" in general.

Whoever the audience, part of the controversy over using animals is the public image of biomedical researchers themselves—how they argue their case to wider constituencies, and how they participate in debate. This is another aspect of the issue, as animal rights organizations attack research and researchers seek to defend themselves; we can learn much about the wider context from analyzing researchers' "public relations" work—issues which we take up in this part of the book.

Public relations specialists manipulate symbols and construct accounts for their clients in order to influence public perception of their clients' actions

(Jackall, 1988). They know how to make expedient decisions appear altruistic, and turn weaknesses into strengths and opponents into demons. This occurs in the context of a diffuse and volatile public, in which everyone is trying to influence their definition of the situation, and reality seems up for grabs.

In this context, many people working with lab animals experience the world outside the laboratory and scientific community as fraught with social risks; as we shall see, there is such a sense of pervasive hostility to animal experimentation that to talk freely or publicly about their work often seems an impossibility. Even the benign attitudes many people have toward veterinarians change once they discover that someone takes care of laboratory animals (Carbone, 2004). In other words, lab workers' accounts portray a situation in which they face social exclusion or are marked by stigma if they reveal what they do.

This pervasive hostility is often perceived as stemming, in large part, from the "propaganda" of the "anti-vivisection movement" (a complex construct, with many nuances). Like other groups facing hostility, the research community has begun to respond, and animal researchers have followed the tactics of other stigmatized groups (Goffman, 1963) in order to win the moral high ground from activists and exclude them from the debate. At the same time, people working with lab animals perceive the "general public" (another multifaceted notion) as latently hostile and open to manipulation because of its ignorance.

There are, then, two broad, though linked, constituencies that the scientific community appears to confront: politically active and publicly vociferous interest groups (the "anti-vivisection movement"), and the critical "general public." These can be treated as "others" against which the scientific community in part defines itself. As we will see, put simply, animal rights activists are represented as anti-human, deceivers, and terrorists, while the general public, by comparison, is viewed as uninformed and too emotional. Set against this, pro-research advocates sought to characterize animal researchers as moderate centrists.

The reason we focus on how these "others" are portrayed is that it is one way that politics, at various levels, is conducted. At the most basic level, "othering" serves to demarcate constituencies with whom it is possible or impossible to debate. In the next two chapters, we will look at how the scientific community does this political work. In chapter 7, we consider how the animal rights lobby is represented in various political processes and events.[1] In chapter 8, we explore how the "general public" comes to be portrayed. In both cases, we trace how "othering" means identifying those groups who have the appropriate qualities for entering into dialogue with the scientific community. Needless to say, what counts as dialogue is itself highly contentious, as we will see.

Finally, in chapter 9, we consider those people addressed by many of these narratives—the general public. How does the general public perceive science, the animal experimentation controversy, and the researchers who use animals? We can answer this by looking at some key findings from various surveys; but we also recognize that "the public" is not a simple, unproblematic entity, but is altogether more subtle. Finally, we emphasize that the complexity of how people view animals is deepened in a context where lab animals are not only strains and models, but, due to genetic innovation, are also hybrids and products. These cultural changes, in turn, have implications for public perceptions of animal experimentation.

VII

Politics, Animal Rights Activism and the Battle for Hearts and Minds

Members of the terrorist arm of the [animal rights] movement have been able to conduct their acts of vandalism and intimidation with impunity because of the disorganized and relatively powerless state of the animal research community . . . [research organizations] are not capable of countering the terrorists, even when their identity is known.

—Nicoll, 1991

If there is a wide gap between researchers and their perceptions of the lay public, there is an unbridgeable chasm between them and animal rights activists. Neither side seems to agree on anything—except perhaps the need to persuade a wider public of their particular case. Spokespeople for animal research must respond to what they see as the threat posed by animal rights activism, trying to counter their allegations; the research community—particularly since the early 1990s, when the quotation above was written—has become more organized to pose political challenges to these threats.

In this chapter, we turn to these arguments, exploring in particular the discourses commonly used by animal researchers to denigrate opponents, and to represent the scientific community in a positive light. What counts as positive is, of course, contentious. As we shall see, pro-research spokespersons use several negative motifs to describe activists (including anti-humanism, violence, irrationalism), while they simultaneously emphasize the putative benefits of animal research. However, such a strategy has a downside in that it presents

133

pro-animal research organizations with uncomfortable dilemmas that might well compromise their arguments.

If animal researchers often see the general public as too gullible or ill-informed, they are much less sanguine about animal rights activists. To some researchers, their opponents do not simply lack proper understanding of science, they are also characteristically misanthropic and bigoted. Describing what they see as the animal rights position, scientists use words like "hypocritical," "fanatical," "hatred of mankind," "intolerant," or "willing to engage in illegal activities" (Paul, 1995). These are the people with whom reasoned discussion is impossible—worse than misguided, they become evil.

Demonizing opponents is an elementary form of exclusion that people in all walks of life use in various ways to establish their own credentials while vilifying someone else's. In the anti-vivisection campaigns of the late nineteenth century, for example, scientists sought to ridicule their opponents as "sentimental, irrational women," referring to the link between that campaign and feminist activities. We discuss here some of the ways that research advocates attempted to exclude opponents, drawing largely on field notes and material collected at scientific conferences and other public forums where scientists were presenting their case for research.

In the course of this work, several themes consistently emerged in the way researchers presented their arguments and the way they portrayed animal activists.[1] In particular, they often picked up on what they considered to be extreme statements by animal rights organizations, That is, spokespeople for research excluded opponents by arguing that their emphasis on animals amounted to being anti-human, and that activists were dishonest and criminal—that is, persons who fall outside the common limits of civilized society. By contrast, research advocates foregrounded scientists' humanity—a rhetoric which seeks both to demonize opponents and to play up medical benefits.

Equating Rats and Humans: Activists as Antihuman

To many scientists, activists' insistence that animals were morally equivalent to humans[2] was antihuman, as it meant that humans would die because animals could not. This was a concern clearly expressed, for example, in one ad the American research community considered running that read: "If animal rights groups have their way, only people will die." Pro-research groups saw this position as so extreme that it disqualified activists from participating in reasonable and rational debate. "Equating the life of a rat or a pig or any other research animal to that of a human is a perversion," an Americans for Medical Progress (AMP) brochure commented. "This insanity must be stopped now."

Researchers expressed outrage over what they saw as activists' immorality regarding humans. For instance, a printed remark by a member of People for the Ethical Treatment of Animals (PETA) that "We are equally responsible to the rabbit dying in a laboratory as to a child dying in the hospital" raised the ire of at least one proponent of animal research (Paris, 1992a), who wrote, "This attitude of the 'animal rights' movement is a noxious doctrine. It boils down to a corruption of human value. The academic term for it is misanthropy—a disdain for humanity." In other literature, Fred Goodwin (1992), a leading research proponent from the National Institutes of Health, personally requested that the pope pronounce the animal rights philosophy as immoral and incompatible with the Judeo-Christian view of humankind.

Almost always, speakers cited at least one quotation by PETA president Ingrid Newkirk to summarize this anti-human position. By far, the most common line attributed to her was "a rat is a pig is a dog is a boy," although cited almost as often was her comparison of the killing of chickens with the Nazi Holocaust: "Six million people died in concentration camps, but six billion broiler chickens will die this year in slaughterhouses." Other Newkirk quotes were also used to underscore the activists' "antihuman absurdities" (Hubbell, 1990:72), such as her charge that humans have "grown like a cancer. We're the biggest blight on the face of the earth," and that if her father had a heart attack, "it would give me no solace at all to know his treatment was first tried on a dog."

Speakers highlighted the absurdity of equating the value of animals' lives with those of humans by showing how animal rights activists did not adhere to this philosophy in their own lives. At conferences and symposia, scientists cited examples of activists being treated for parasites and struggling over whether shrimp and insects had rights—suggesting that the activists did not hold all animals to be unequivocally equal to humans. One speaker interpreted Newkirk's quote as meaning that she held animals to be more important than humans. "I think there's something noteworthy in the fact that the first one she mentioned was a rat," he told an audience of about 100 people. He went on to say that animal advocates were misanthropes and that the implications of their views supported the suffering of people:

> One of their slogans, one of their most popular ones that they chant at almost every demonstration, is "Stop the pain and stop the torture." I think that's something that everyone here holds some sympathy for. But the real meaning behind that is to stop all research, no matter what the pain and the torture to our fellow human beings is. . . . The president of Last Chance for Animals says, "A life is a life. If the death of one

> rat cured all the diseases, it wouldn't make any difference to me. In the
> scheme of life, we're equal." One rat, all diseases. That's not stopping the
> pain and stopping the torture. That's continuing the pain and continuing
> the torture!

According to research advocates, the "anti-human absurdity" rested on extreme boundary-blurring by activists. To demonstrate the animal rights perspective, one seminar speaker showed a cartoon that had a rodent at the top of a pedestal indicating different life forms. He also showed a slide of someone dressed in a Mickey Mouse costume and said that Mickey was "the quintessential animal that was not an animal but was seen and treated just like us"; this kind of thinking about animals was, he said, now common to many people.

At the same time that pro-research organizations pointed out the logical and moral problems with the activists' boundary-blurring and anti-human sentiment, they drew clear boundaries between animals and humans by affirming animals' scientific purpose. One organization, for example, distributed buttons and pamphlets with a picture of a white rat on them: "Some People See a Rat, We See a Cure for Cancer." The humanity of researchers is thus clearly foregrounded.

Outrageous Distortions: Activists as Dishonest

A second way of demonizing activists was to suggest they are not honest. One AMA speaker observed, "The worst aspect of the movement, besides lab destruction, is the absolute dishonesty and the outrageous distortions these people engage in." Once they were seen as dishonest, animal activists could hardly be trusted to engage in genuine debate and should therefore not be permitted to participate. In a speech delivered to the Research Defence Society (of Britain), Goodwin (1992) of the National Institutes of Health noted: "[The] 'stop research' strategy employed by the animal rights movement is the dissemination of falsehoods regarding animal research. And, as de Tocqueville noted 'The public will believe a simple lie rather than the complex truth.' [A] major problem for the scientific community is that the animal rights groups have been telling simple lies, and we've been responding with complex truths."

Such "simple lies" included out-of-date or sensationalist images characterizing researchers as torturers of animals. At one conference, for example, a pro-research speaker showed a slide of a PETA ad reading, "If you like pulling off wings of a fly you might like a career in biomedical research." Another suggested that images, such as an anti-vivisection ad featuring an apparently uncomfortable monkey struggling in a stereotactic braces, were "staged" by activists.

Other "simple lies" entailed allegations that opponents falsified the actual nature of experiments and their results. Thus, one speaker at an AMA workshop, objecting to being "made out to be a liar" by an animal rights activist, went on to criticize the writing of Brandon Reines, who allegedly claimed that Harvey's heart research in the seventeenth century did not involve the use of animals in his experiments. He argued that Reines selectively used Harvey's writings and left out explicit references by Harvey to the use of animals. "He falsely represented Harvey's work. This is what you're up against—our opponent is shameless in its presentations."[3]

Activists were also accused of pseudoscience and of cloaking their feelings to make their views more acceptable. Thus, Goodwin (1992) observed that because of their "inherent dishonesty," activists could deflect public discussion away from what in his opinion was the key issue: the animal rights philosophy of animals and humans as morally equivalent. Instead, they engaged in "pseudo-medical or pseudo-scientific arguments." When the physician-founder of the Physicians Committee for Responsible Medicine, the "scientific apologist for PETA," talks about any scientific aspect of animal research, the credibility of his remarks must be questioned, suggested Goodwin. In his words: "I object when an activist attempts to hide his or her true position by professing expertise on research, or on the history of medicine, or on the way animal research has or has not contributed to human health."[4]

Similar tensions between activists protesting the moral basis of science and scientists objecting to "pseudoscience" on the part of opponents occur in many areas of scientific controversy (Nelkin, 1992). They are part of a wider cultural mistrust of science in general, expressed over issues such as environmental damage, nuclear fuels, and trials of genetically manipulated crops, creating a resistance to what people perceive as the ways scientists seem to see the world only in terms of its use value (Jasper and Nelkin, 1992). Yet only a few of the researchers we interviewed made reference to these wider issues and other concerns; those few who did suggested that animal rights was part of a broader "green" movement, to which the public (and particularly school children) were increasingly exposed. Most researchers, however, made no mention of these various other ways in which the public might perceive them as doing wrong; rather they focused very specifically on the way that animal rights activism might damage medical progress.

Aiming animal rights arguments at impressionable audiences was held to be particularly pernicious, at least partly because of the apparent success of the green movement in appealing to young people. Using language reminiscent of the claims made by activists about researchers' fascism, one speaker referred to activists' "propaganda" in schools, saying that it was a "characteristic of fascism to go to the young. It was true of the Nazis, and it's true of the animal rights movement. . . . I am struck by the number of young people who think that science has made the world a worse place. They are disillusioned, one step away from needing treatment" (referring to mental health care). More than just depicting science as harmful or useless, this "propaganda" portrayed scientists as "evil," warned Goodwin (1992): "In the United States, the bulk of the new investments that PETA is now making is in the schools: Students for Animal Rights, PETA kids. Animal rights music festivals. In Montgomery County, Maryland, where large numbers of the children of NIH scientists are enrolled, the PETA people get into the school under the guise of teaching about pet care. But the real agenda is to portray biomedical scientists as evil."

Pro-research advocates seemed to see students as innocents being led astray by animal rights activists' simplistic and emotionally appealing lines. The president of Americans for Medical Progress (Paris, 1992b), for example, observed that "activists shrewdly manipulate audiences with non-threatening, seemingly logical arguments. Teachers, students and parents are sucked in by their honey-sweet lines and fail to recognize the swarming hive they are stepping into." Paris (1992/1993) also noted:

> Schools and our children have increasingly become part of the political battlefield . . . they [animal rights activists] have launched a planned effort to get to our children before they are old enough to understand the issue, and before they might pursue careers in the biomedical sciences. Children too young to fully understand the moral distinction between humans and animals are returning home from school as vegetarians (which can be physically dangerous for children) and opposed to biomedical research. . . .
> I encourage you to check your local schools to make sure their curricula have not been corrupted by the animal rights activist agenda.

In addition to cloaking their ideological agenda, animal rights activists cloaked their organizational identities, according to pro-research groups. One of the most widely distributed pieces of pro-research literature was an article written by a journalist, Katie McCabe, (1990) for *The Washingtonian*. Besides alleging that one of PETA's founders staged photographs of animal cruelty, she presented pictures of PETA leaders in formal dress at a fund-raiser above the caption: "The Glitterization of Animal Rights." This was part of a boxed section that claimed that PETA used its funds to pay the legal fees for Animal Liberation Front (ALF) members who had been convicted of university break-ins, rather than its stated goals of education and outreach.

Practicing Terrorism: Activists as Criminal

Violence was, not surprisingly, a concern of pro-science spokespersons. According to one AMP brochure, "The more extreme animal rights activists are resorting to terrorism: fire-bombing, vandalism, threats, slander, burglary, physical harassment, and even attempted murder in their effort to hobble the advance of medical science and the saving of human lives." At an AMA workshop, a speaker warned that "science as an enterprise is at risk due to scientific terrorism. Activists' destructive tactics threaten the future of research." In such accounts, animal rights activists fell outside the boundaries of civilized society and could be given no place at the table with reasonable people.

Sometimes, pro-research presentations used simple statistical summaries to show the number of institutions targeted or broken into. One seminar speaker in the U.S. thus observed that 54 medical schools have been told they were targeted and almost 4,000 incidents have occurred at medical schools and researchers' homes. At other times, speakers used visual evidence of break-ins as part of a slide show or a video. Slides shown by one speaker included, for example, a newspaper article about an ALF arson attack at a University of California, Davis lab that caused $4 million damage; a picture of a wall at a Northwestern University lab—"a graffiti attack"—saying "animal researchers are corrupt and perverse," and "animal researchers are Nazis"; a Louisiana State University lab wall sprayed with "Cat killer, go to hell" because the researcher, who had regularly received numerous threats, was "under attack"; and a postcard received by this researcher with a photo of him with a skull drawn over his face and a death threat written under it. Others depict still photographs of swastikas painted on lab walls by activists.

The violence often goes beyond graffiti and threats. Thus, speakers in an animal training workshop at a government research institute in North Carolina showed their trainees a video showing clips of burning buildings and fur-retailing stores and animal research laboratories in London after attacks by animal rights activists. The footage showed journalists interviewing an ALF member in what they described as "a seedy hotel." The interviewee's face was masked with a black stocking, and his voice was disguised to preserve his anonymity. This video, according to the speaker at the workshop, "adequately reflects the kind of environment in which we find ourselves using animals in research today." On other occasions, pro-research advocates simply quoted activists as proof of their terrorist intent. One AMP brochure, for instance, cited Tim Daley of the ALF, who purportedly said: "In a war, you have to take up arms and people will get killed. . . . I can support that kind of action by petrol bombing, bombs under cars, and probably, at a later stage, shooting [researchers] . . . on their doorsteps."

Well-known scientists who have been targeted sometimes speak out about their experiences. One celebrity victim, for instance, focused on the four times he had received bomb or death threats. Another researcher talked about how bomb threats "had a devastating impact on my life," making him live in fear for himself and his family. In some of these anecdotes, threats were made against scientists not just because of the nature of their research but also because they had spoken out against activists. Speakers acknowledged that many of their peers were uncomfortable talking at public functions because they felt like they were not trained to speak in this capacity and because they were afraid that such speaking would make them targets. At one seminar, a researcher said that "I'm here as the designated attackee. I was attacked. I was attacked, not just because of my research, but because I had been speaking out on behalf of animal research, and noting the dangers posed by the animal rights movement. . . . ALF was in my office one month later. They looked for proof that my speaking was supported by the government, but they couldn't find any link. But I am now working for Fred Goodwin. I get the usual death threats and hate mail."

Relatively few scientists are as willing to "come out" so publicly; most researchers do not seek to present their case for research in the kind of public forums we have discussed here. Yet very much the same themes occurred in our less public interviews with researchers in the U.S. and the U.K.—the anti-humanity of activists, their dishonesty, the effect of the green movement, the risk of terrorism. Over and over, we encountered the same anxieties about what the animal rights activists might do.

By labeling activists as dangerous, pro-research groups have rallied resistance, however. They have, for instance, pushed to legally classify laboratory break-ins as terrorist acts and have organized—and made public—victim support groups. A politically conservative consulting firm produced a report (Hardy, 1990), supported and distributed by the AMP, which catalogued the "terrorist" activities of animal rights groups and attacked the reticence of federal authorities to enact legislation to deal with such terrorism. This report, like the firm's other publications, was written for court decisions to use and cite. Similarly, a more recent government paper produced by the British Home Office and Department of Trade and Industry spoke of the need to protect research against extremists' actions, noting that "Some of the activities of animal rights extremists are often considered to be acts of terrorism" and pointing out that these actions are covered by the 2001 Terrorism Act (Home Office, 2004).

Another response has been the creation of Victims of Animal Rights Extremists (VARE) in Britain, which has made public the harassment of people attacked for their involvement in research or breeding of laboratory animals. Their website includes victims' stories: "Meg," a manager in contract research, for example, described

> an explosion outside the house in which [cars were] written off and there was £5,000 of damage to the front of the house, including the window of the bedroom where we were sleeping. It did affect us dreadfully. . . . For a while I could not go out of the house. You feel very vulnerable every time you leave the house and worry that someone is following you. . . . Each time there are animal rights demonstrations in the region, you worry whether they will come again that night. I am doing an ordinary job in a legitimate industry and I am being threatened purely because of this. It should not be allowed.

Pro-research groups, then, contest the legality of their opponents' activities, especially when these take the form of overt attack. They might then use this to address politicians directly, appealing to them to change the law (with evident success in the U.K., where the Home Office website has posted information about the British government's desire to legislate against attacks specifically by animal rights activists).[5]

More generally, by representing opponents as irrational or criminal, pro-research advocates are attempting to set the terms of the debate, willing to engage in discussion only with those deemed rational, honest, and nonviolent. Advocates thus excluded most animal rights activists from participating because they would not be orderly and civilized. Animal rights people, too, use such tactics, separating themselves from the extremists who use violent action; speaking about the "very small number of people . . . who engage in illegal activities," Wendy Higgins, campaigns director at the British Union for the Abolition of Vivisection, complained that "inevitably groups like VARE [Victims of Animal Rights Extremism] paint a picture of the entire movement as a bunch of extremists" (quoted in Bhattacharya, 2004).[6]

Whatever their views on the rights and wrongs of using animals in experiments, the two sides concur in painting a picture of opponents as irrational others. To speak of others in this way positions the speaker as more moderate, and hence more likely to be acceptable to a wider audience. Trying to address this wider public, scientists not surprisingly portrayed themselves as reasonable, as well as emphasizing their own expertise in looking after animals and the importance of the research to medical outcomes. Moreover, in the face of the threat of violence, they often sought to express their work as heroic.

Strategies of Defense: Occupying the Middle Ground

To posit themselves as a party willing to debate in a reasonable manner, re-search advocates sought to define the middle ground. By finding quotes from activists that called for "total abolition" of all animal experiments, the pro-research community (e.g., Horton, 1989:742) described how the animal rights movement represented a radical position. As one speaker remarked, reiterat-ing Ingrid Newkirk's words, "the animal rights movement is not interested in bigger cages but abolishing all cages." Real dialogue, research proponents argued, could not take place with those who sought such total abolition, but only with those critics who were more moderate.

Pro-research groups frequently cited attitude surveys of the public and the medical profession to buttress their case for centrism, suggesting that the research community was not a minority voice, let alone an extreme one. For example, literature produced by Americans for Medical Progress, along with other pro-research literature and presentations, frequently made reference to a survey claiming that 77% of those polled agreed that the use of animals in research was necessary for medical progress. And the AMA (1992) claimed that "polls show that the American public overwhelmingly rejects the activ-ists' claim that there is no difference between animals and humans? So, where does that leave the animal rights movement?"

If the animal rights movement is one extreme, the research commu-nity, proponents argue, now occupies the center. In this regard one confer-ence speaker noted, "the research community has moved very definitely into the middle ground" because it advocates neither "an absolute commitment to the animal's comfort and procedural safeguards," nor a blatant disregard of the well-being or suffering of animals. This loosely positioned the biomedical research community with traditional animal welfare organizations. Similarly, "As scientists, we can consider ourselves as protectionists. We care for ani-mals," an NIH speaker noted.

Yet to be positioned with animal protection groups is a delicate mat-ter, needing the good to be sorted from the bad. Literature from the research community frequently described how the old traditional animal welfare or-ganizations, which truly represented public opinion, were being taken over and radicalized by the leaders of the animal rights movement, who were con-sequently acquiring the control of larger sums of power and money. Referring to the ASPCA, for example, Goodwin (1992) of the NIH observed that it had shifted toward the animal rights end of the spectrum, "but, thankfully, many local community charities in the United States remain true to the principles of animal welfare, and have not been corrupted by animal rights groups." This advocate went on to say that "In the past, the ineffectual response of the scientific community and others to the animal rights challenge contributed to a gradual radicalization and takeover of many traditional animal welfare organizations. It is no longer easy to determine whether some humane societies are in the welfare position or the rights position."

Advocates are under no false illusions: people who adopt the "rights" position are unlikely to be won over. Although they primarily blame the public for its inability to enter into dialogue, they also faulted their own community. Researchers have not, several told us, done a good enough job in reversing negative images of animal researchers and their work, nor have they been forceful enough in putting across the case that animal research saves lives. Only by emphasizing the benefits, they felt, could they effectively put up a counter-argument to the emotions invoked by animal rights extremists.

Researchers seek to emphasize the middle ground not only with regard to treatment of animals, but also by detailing medical progress. Pro-research organizations put enormous stress on how animal research has been basic to the conquest of disease, typically drawing attention to how people have been saved from particularly awful illnesses. By taking this triumphal stance, they not only defend their moral position but also seek to persuade a sceptical public of their role. More than merely extolling the benefits of research, these accounts characterized investigators as unsung heroes whose future achievements were jeopardized by the growing voice of animal rights activists—rather like the portrayal of lab animals as unsung heroes that we noted in chapter 3. Pro-research speakers would, for instance, often begin their presentations by asserting that medical research has added twenty-six years to our lifespan this century. The first pages of a prepared speech for the public (AMA, 1992) noted:

You've heard people wish for "the good old days." We all know what they mean by it, of course, but the fact is, the good old days weren't all that great for most people. For example, the average life expectancy of a person born in ancient Greece or Rome was about 20 years. A couple of thousand years later, things still hadn't improved much. In 1850, the average life expectancy was about 40 years—at least in places like Massachusetts, which was probably one of the more healthful areas of the world. In another 50 years, by 1900, the average lifespan of a person born in this country had crept up to 47 years. What about now? The average American born in 1990 will live nearly 76 years—and enjoy a healthier life than his or her ancestors. It's easy to forget how much our medical well-being has improved.

Conference presentations, like most pro-research publications, similarly litanized "one-time scourges" purportedly defeated through animal research. These usually included anthrax, cholera, smallpox, measles, tuberculosis, leprosy, tetanus, diphtheria, rabies, whooping cough, and hepatitis B, as well as some diseases that were "brought under control so long ago that their names are now quaint and unfamiliar," as in the case of pellagra, rickets, or beriberi.[7] Spokespeople agreed that the problem was that medical achievements had been taken for granted. A surgeon at an American symposium described his experience of practicing in Third World countries as "a glimpse into the past" without biomedical research. "Flocks of children," he told the audience, "regularly die of smallpox and measles." Another surgeon even admonished his own colleagues for forgetting medicine's contributions:

> We do not appreciate how far we have advanced in the last century, and how important animal models have been in achieving these advances. Who among us has ever seen anyone die of whooping cough, measles or polio? We take for granted these vaccines that were developed with animals and enabled these life-threatening infectious diseases to be brought under control. Most current practicing physicians don't know how lethal common bacterial infections were prior to the advent of penicillin and of many generations of antibiotics that have since been developed and tested in animals before administered to humans. We've forgotten that people in our grandparents' generation frequently died from complications of appendicitis and communicably acquired pneumonias.

Polio received particular attention, perhaps because it lends itself more readily to visual images, such as a photograph of over fifty iron lungs, included in a brochure put out by a research-based pharmaceutical company in North Carolina. Such a photograph, supplied by the March of Dimes Birth Defects Foundation, might have been used in its day as a symbol for the conquest of medical technology in preserving human lives. But what it does in the brochure is symbolize the miserable conditions of human existence prior to polio vaccines developed through animal research. "In the 1940s and '50s," the caption read, "polio wards such as this one were common place in the United States, where 33,000 polio cases were reported in 1950 alone. Animal research was vital to the development of vaccines that brought the disease under control in the 1960s." The conquest of polio is such an important symbol of researchers' triumphs that the Coalition For Animals and Animal Research (CFAAR) used a life-size iron lung to counter an animal rights demonstration. One of this organization's newsletters (Russell, 1990) described how participants celebrated the anniversary of the introduction of the Salk polio vaccine by rolling the lung into a shopping plaza where animal rights activists were staging a protest against research:

Rolling over the concrete toward the Plaza, gaining momentum, the iron lung made quite a dreadful noise and attracted considerable attention . . . a young woman asked a CFAAR demonstrator what gave people the right to use animals. "What makes us special?" she asked, in a voice that clearly indicated she thought the answer was "nothing." The CFAAR member turned to the iron lung and declared, "This is what makes us special. Only humans could invent a device like this, that saved hundreds of thousands of lives, and only humans could have developed a vaccine to conquer polio, to make the device unnecessary."

While these success stories made researchers look like heroes who worked "miracles," this image was more easily sustained by relying largely on past successes, such as the smallpox or polio vaccine. Because there were fewer recent "magical bullets" to cite, advocates relied far less often on recent biomedical research, described by some as "half-way technologies" (Thomas, 1977) because of their expense and failure to provide a true prevention or cure.[8]

Pro-research advocates dealt with this problem by suggesting that scientists were "on the verge" of achieving breakthroughs with many major diseases. For example, AMP's newsletter, *Breakthrough,* was devoted to publicizing the advances made by researchers for problems such as ulcers, migraines, cystic fibrosis, cancer, arthritis, and Alzheimer's disease. Reports typically focused on small, although necessary, gains in basic research that could lead one day to actual cures. "Breakthrough" is a potent motif in biomedical research in general, not least insofar as it is a means to gaining reputation or attracting venture capital. As Brown (2000) notes, breakthrough announcements work by privileging certain features over others. Brown illustrates this by showing how a transgenic "breakthrough" in xenotransplantation might indeed combat acute rejection, but that such announcements function by drawing attention away from the various other sorts of rejection that are yet to be addressed. Similarly, breakthroughs in various research programs entailing animal experimentation can be seen to detract from other, potentially more effective public health initiatives.

Pro-research advocates dealt with the issues surrounding the actuality of biomedical advance by intertwining discoveries of vaccines and cures with diagnostic advancements and even current research, thereby blurring the line between magic bullets, half-way technologies, and the promise of experiments. Not surprisingly, their critics are well aware of this problem and frequently point out the gap between the promise of medical cures and the realities of research.

"Surely They're Worth a Few Laboratory Animals?"

Patients themselves are often relied upon to provide testimony, and to argue that they would not have survived if it had not been for the deaths of some animals. One man, representing the organization Incurably Ill for Animal Research (IIFAR), explained at a pro-research symposium how he had been severely burned in a plane crash and treated with surgical techniques and drugs that he attributed to animal research. During his presentation, he made a point of thanking the research community:

> Yes, the benefits of research are very real. And to some of us they are evident every day. And we're grateful for the research that's been done in the past that's helped us. We're grateful for the researchers and the rest of the research team that has made all of research possible. And I know that we have some members of the research community here with us tonight. And I'd like to take this opportunity to say thank you for myself personally, and the rest of our members. 'Cause I think that's something that's not said up 'til now.

Not only do individual people offer appreciation, but an entire pro-research organization was formed in the U.S. with the name "Thank You, Researchers." More than merely thanking researchers, their conference speeches told, almost in a religious sense, of how they had been "blessed" by the researchers who had worked "miracles" by making it possible for people like themselves to be "reborn." Who could question the loss of a few animals, they argued, given the benefits of their research. To wit, a woman who had received both a liver transplant and a hip replacement spoke at one seminar about her experience and gratitude to researchers:

> I have been blessed by every member of this room. I am extremely grateful to all of you. We must keep in mind the main thing—biomedical research saves human life. I had primary biliary disease. I was told I would die in two years, but I'm still here . . . because I have benefited from scientific investigation involving the use of animals. I watched my son graduate from law school, I celebrated my 35th anniversary, but 10 years ago I lay comatose and dying. My family was told to plan my funeral, write my obituary. [She then tells a story about how she acquired someone's liver]. It saved my life. . . . How can I describe what it feels like to be reborn? How can I explain to you how grateful I am to you for each new day? How truly terrified I am that your vital research will be deterred by misguided animal rights activists and that other people will be deprived of their rebirth. Those who impede medical research do a disservice to themselves. We cannot afford to put one medical miracle at risk.

These testimonials, as well as presentations by investigators, frequently ended

on notes of both optimism and pessimism, weighing medical developments—
"miracles"—against the risk of closing down the labs. While presenters were
generally optimistic about the possibility of developing treatments for disease
through animal research in the future, they also added an apocalyptic warning
that because of a "misguided," "overzealous," and "fanatic minority" we risked
"throwing it all away" and "depriving our children" of the benefits of such research.[9] Thus, these heroic narratives posited the animal rights movement as
the primary obstacle to new medical miracles. It was these miracles that would
be in "danger" and "at stake" (AMA, 1992) if the biomedical research community's voice did not prevail in the animal experimentation controversy.

Heroism—helping patients, particularly children—is a consistent theme
in these narratives. In part, this image helps to counter animal rights accusations that researchers undertake unnecessary research to promote their own
careers; but it also serves in part to soften the public's view of scientists as cold,
insensitive intellectuals. Instead, they become rescuers, coming to save victims,
usually shown as healthy and appealing young people and animals, from the
ravages of disease. Thus, the pro-research community seeks to counter what
they see as the emotionalism of animal rights publicity by themselves using
highly emotive imagery.

The most emotive images feature children: if the research stops, children
would be victims. One AMP ad in *The New York Times* and *The Washington
Post* ran a photograph of a smiling, attractive, healthy-appearing young girl
above the caption "She May Owe Her Life to a Rat, a Monkey, and a Lot
of Dedicated Scientists." The text noted: "There's good news for the 30,000
young victims of Cystic Fibrosis. They can now hope to live a long time . . .
[this] dreaded genetic disease . . . may soon go the way of diphtheria, smallpox and polio . . ." Similarly, a poster put out by the Association for Research
in Vision and Ophthalmology in the U.S. showed an apparently healthy infant, its head being cradled in human hands. "Her Eye Disease Comes From
Birth. Her Hope, From Animal Research," the bold print read, while a poster
for the Foundation for Biomedical Research asked "If We Stop Animal Research, Who'll Stop the Real Killers?" Beneath the large print, three en-
larged microscopic images were shown of cancer cells, diseased arteries, and
the AIDS virus.

Such rhetorical devices again preclude animal rights people from the
controversy because they shift attention away from philosophical arguments
about the value of animals' lives and accusations of researchers' cruelty to the
positive outcomes of the research itself. Such narratives focus on the ends
rather than the means of animal research. The ethical choice was not whether

or not to use animals in research, but whether or not to deprive those who are sick and vulnerable of its fruits. "Is it ethical to let a child die of a disease when you know that animal research can save her?" an IIFAR speaker asked rhetorically at a pro-research symposium.

Members of CFAAR in California thus wrote in a letter to the editors of *Science* (Denver et al., 1988): "We have learned that when we are dealing with the activists and with the public, the most effective approach is not only to use rational dialogue, but also to make the topic emotionally appealing. We therefore show photographs of young, healthy children, healthy, elderly persons, healthy cats and dogs, and include a brief statement about how animal research benefited them." They acknowledged, however, that most researchers would find this kind of emotional response difficult. As the NIH's Goodwin (1992) commented:

> . . . it is difficult to get academic researchers to over-simplify and to infuse their arguments with emotion. Goes against our grain. We tend to be logical, reflective, and careful, and not to overstate anything. But in a political battle with a radical group, a purely intellectual approach is ineffective and even counter-productive. . . . We have to develop brief, emotionally engaging examples of the medical benefits of animal research, in terms understandable to the public. Medical research is an esoteric, complicated field, and most scientists are not particularly good at public education; that is not why they went into science.

Similarly, two public relations experts told an audience of scientists, physicians, and technicians at an AMA workshop to "talk about your research to people with the same passion you talk about your patient care." Following this advice, physicians invited to speak on behalf of animal research began their presentations at symposia by saying that they were going to share "personal interfaces" or "share some stories." *The Compassionate Quest*, a brochure distributed by the North Carolina Association for Biomedical Research, quoted a research physician as saying: "When one of my patients dies, all I can do sometimes is come back to my office and cry. It's where the issues of animal studies gets very clear for me. I don't like to see an animal die. But I hate to see a child die."

Children undoubtedly add extra emotional appeal to campaigns. Slides and color pictures frequently depict physicians bent over or helping children, with accompanying narratives describing how a child on the verge of death has been brought back to life using techniques developed from animal research. One brochure thus told of a child called Charlotte, born with "spaghetti-thin arteries in her lung." It was "a race against time" until doctors perfected the heart and lung transplant and the drugs necessary for her survival. Eventually, Charlotte's situation was ameliorated when, after performing the technique on lambs, a doctor performed a balloon angioplasty on her pulmonary a ies. At a symposium, the speaker representing IIFAR played a video ot ne₁ mother explaining: "My daughter was in that doctor's hands, and he brought her back. Surely it's worth a few laboratory animals."

New Dilemmas and Research Advocacy?

Researchers have responded to the emotional arguments about animals by emphasizing their humanity and benevolence in bringing about the benefits of research—the "miracles." In doing so, they attempt to show the public that they have not lost touch with their emotions and are thus sensitive, competent participants in the debate—they care about children and animals, too. The "real killers," they suggested, were the diseases that the researchers fought against.

At the same time, pro-research accounts implied that animal rights activists were not aware of the hardship of human illness, and therefore their preoccupation with animal suffering was misguided. Lost in heady philosophical ideas, they did not inhabit the real world and therefore did not appreciate the difficult choices that medical scientists had to make. "It must be nice," a veterinarian in a medical school told us, referring to the animal rights philosophers, "to have chosen a profession that is totally dealing with words and ideas. I think all you have to do is see somebody in the burn center or watch your wife go through an appendectomy as an adult and have the surgeon tell her a fair number of adults die from appendicitis to feel a commitment to humanity and a little less strong commitment to animals."

Pro-research advocates thus use various rhetorical means—emotional and rational—to get across their argument, to make a counter-attack. They present themselves as moderate centrists, experts, and heroes, while portraying animal rights activists as anti-human, dishonest, and criminals. Such tactics are familiar, characterizing many other controversies centered on moral issues (fetal research, for example; see Maynard-Moody, 1992). By doing so, they have sought to capture the moral high ground in ethical decisions by promoting a moral identity that is superior to their opponents, and have attempted to exclude or restrict the voice of these "others" in the controversy over the use of animals in biomedical research.

How successful the pro-research movement will be in winning influence over decisions surrounding animal experimentation remains to be seen. But if nothing else, their narratives could create a new identity among researchers that has been lacking. At a time when science is frequently under attack and vivisection is once again morally tainted, biomedical research advocacy can convert uneasiness into confidence. One young graduate student, for example, described her local CFAAR chapter as a "support group." She went on to tell us: "Before we formed there was really no group on campus that was pro-research. We assume that most of the scientific community is pro-research, but they don't have any formal group to say that they are."

Yet pro-research organizations present researchers with new dilemmas. Scientific authority rests upon clear boundaries that distinguish scientists from the laity (Gieryn, 1983). Such boundaries have traditionally sought to demarcate scientists as neutral rather than judgmental, rational rather than emotional, and autonomous rather than malleable to popular opinion. It is precisely these boundaries that researchers and educators demand when dealing with, for example, the emotional reaction of students to dissection. But if what distinguishes animal researchers from animal rights activists and the public is autonomy and an unemotional persona, then the formation of interest groups and humanistic and emotional presentations can blur these distinctions. New boundaries may have to be created. Indeed, animal rights activists have exploited this blurring as an opportunity to label the researchers as the emotional and insidious party in the controversy.

At a pro-research rally, for instance, we observed animal rights activists staging a counter-protest. One activist insisted, referring to the pro-research literature, that the scientists were the harbingers of sentimentality and misinformation in the controversy: "It always amazes me that our opponents accuse us of being overly sentimental, of being misinformed . . . yet the three pictures that they are always working with, and that you see here today, are, in my humble opinion, classic examples of mushy sentimentality and distorted propaganda!" Here, then, there is a mirror image of rhetorics.[10] On the pro-animal research side, numerical improvements in health brought about by animal experimentation and individual human suffering are stressed; on the animal rights side, the numbers of animal deaths are emphasized and the suffering of individual animals highlighted. Each side can thus accuse the other of sentimentality (as well as problematize the other's "facts and figures").

Each side thus ups the stakes. Both produce arguments and counter-arguments, reason and "mushy sentimentality." In this chapter, we have focused on the pro-research lobby, and how it has responded to and represented the animal rights lobby in its public pronouncements. We have described a number of forms of argument used either to denigrate animal rights or to elevate the research community. As we have also seen, the strategies used raise problems of their own. In particular, we have shown how the rhetorical contrast between rationality and emotionality used by pro-research advocates is frequently breached by recourse to the "human-interest stories" of patients. As such, while on one level, this boundary is constantly policed (with both sides laying claim to the better "rationality"), simultaneously emotionally charged representations abound on both sides.

As with any subculture, not everyone shares in it (Warren, 1974) or wants to stand on the firing line. Some, as we noted, withdrew from the controversy and felt stigmatized by public reactions, a theme we will explore in the next chapter. A few of our informants described anguish about having little experience in appearing on television and concern that the academic community was being dragged into "the street corner" along with the animal rights activists.

By contrast, there are some research advocates who positively enjoyed the challenge of debating activists and expressed dissatisfaction with colleagues who failed to do so. One of our informants, a researcher-physician, was the son of a preacher. He told us about how he enjoyed making public presentations and viewed challenging animal rights activists, along with anti-abortion activists, as "a sport":

> I used to make a great sport of walking into the hospital through the front door so that I could confront one of these individuals [anti-abortion activists] and look them in the eye and say: "tell me, how many adopted children do you have?" I have never met one who had an adopted child. And I feel much the same way about these people [animal rights activists] that I characterize as "humaniacs."

As sociologist Erving Goffman (1963:27) observed with respect to people who make careers out of what might be perceived of as their stigma: "instead of leaning on their crutch, they get to play golf with it." Winning hearts and minds—persuading the public that animal research is necessary—means getting out of the bunker.

VIII

Rationality, Stigma and the "General Public"

> Do we have to rethink . . . what we're doing completely
> because of public opinion, or is it just that public opinion
> has been swayed in a direction and if we put in a different
> viewpoint we'll sway them back? . . . my only way forward
> is, we've got to actually stick our heads above the parapets
> to show we are actually normal ordinary people trying to
> do a good job for mankind.
>
> —British scientist, interview

Even though some researchers are willing to "stick their heads above the
parapets" and speak openly to the public (or even to challenge animal rights
activists), many more were not. On the contrary, the reference to parapets was
typical of several who spoke about feeling besieged by hostile public opinion.
Yet they must—as the quotation above illustrates—attempt to sway the public.
One line of defense is to position themselves as moderate centrists, as we have
seen, while attempting to define who might be potential allies. In turn, this
means defining those who are outside any possible alliance, who are unlikely
to be persuaded by pro-research arguments.

Yet however much researchers may feel that their research and its objectives are ethical, they cannot always persuade outsiders—especially those who
do not work in science. It is in these situations particularly that researchers may
feel besieged or stigmatized, and they need to find ways of addressing potentially hostile audiences. In this chapter, we consider scientists' sense that they
are confronted with a wider public skepticism, and sometimes hostility, and

explore the tactics they use in dealing with this—tactics that include routine concealment (and the warranting of concealment not least by portraying the general public as "others"). But, as we shall see, this also includes the discursive definition of a public with whom it is possible to conduct dialogue and be potential allies.

Dealing with Stigma

Despite efforts to be positive, most researchers are leery of discussing too publicly the work they do; they are too aware of public disapproval. Goffman (1963) suggested that people experience stigma when they possess certain qualities that lead to negative social consequences—through exclusion or anxiety, or through discrimination and social disenfranchisement. In the face of this, individuals inevitably seek ways to neutralize or avoid further stigma, including trying to present counteracting images of themselves. Thus, owners of stigmatized dog breeds such as pit bulls try to present themselves (and their dogs) in a favorable light, or deny that the dogs are even pit bulls (Twining, Arluke and Patronek, 2000). Until recently, studies of people in stigmatized work focused on those in marginal occupations. But because of shifts in public perception, some forms of work are now stigmatized that once had merited greater prestige. People in these occupations might now find themselves the target of moral crusades by groups seeking to change public opinion about whatever they find offensive—such as doctors performing abortions, or engineers building nuclear plants. Lurking behind such moral criticism are often implicit charges that these workers must be unprincipled or shameless to do what they do.

Yet to the targets of these crusades, stigmatization is often puzzling, especially when their profession once had higher standing. They may react to this by going into the closet, to conceal who they are. Such strategies of information control enable workers to deal with others and provide "recipes" for an appropriate attitude toward the self (Goffman, 1963). Having an occupational stigma, however, does not follow automatically from doing the job; rather, it is relevant to self only if it is perceived as such by the actor, which in turn depends upon that person's experience with public rejection.

Public rejection, we were told by many lab workers in our research, was a common experience. Part of the stigma has to do with the gap between what is acceptable practice toward animals outside of labs and what can be justified inside them. But another part has to do with the gap between lay and scientific beliefs about science itself. One important insight emerging from sociological studies of science in recent years has been the observation that scientific knowledge is extraordinary only in that we believe it so; other than that, science is just another set of human practices. Yet scientists routinely go out of their way, in their use of language, to present science as authoritative knowledge, with the mundane messiness of daily laboratory life obscured (as we saw in chapter 2). Studies of the public understanding of science, furthermore, have emphasized public disquiet about the presentation of science as authoritative or certain; that is, people seem increasingly to question whether it is credible for scientific institutions to make statements of unmitigated fact. So one source of stigma has its roots in the way that many laypeople feel that science is much less certain, and more morally dubious, than its practitioners sometimes claim. This is a stigma attached to many areas of science—genetic engineering or cloning, for example—which concern the public. Here, too, scientists might privately feel similar anxieties about the ethical dilemmas but maintain their belief in public that science is overridingly good.

Inevitably people working with lab animals often feel that they cannot readily admit to what they do, or even that they feel besieged—stuck "behind the barricades," as several scientists described it. Both researchers and technicians reported how they believed others saw them—as, for example, "cold and detached," or not even human. One investigator explained how she and her peers were seen as "barbaric" because of their use of animals. Animal technicians are particularly vulnerable to stigmatization by others—they cannot, after all, easily skirt the issue, since animal experimentation is their entire job, and they sometimes have to transport animals through public spaces. Not surprisingly, technicians often spoke about their reluctance to admit in public what they did, for fear of being labeled "mouse murderers" or similar. One admitted, for example, that she was reticent to say what she did even to her own daughters who were "into animals": "I tell them I work at the University," she said.

Lab workers quickly learn to be reticent about describing what they do. One technician explained how he "was in line at a grocery store . . . and I started talking with the woman in line behind me. She asked me what I did, and I told her. Her immediate reply was, 'You should be ashamed!'" Others reported the difficulties of moving animals around, and how they often put a cloth over the cages, or moved animals in boxes, to avoid looks of "disgust . . . like we're the devil or something," or comments such as "You're not one of those people who experiment on animals are you?"

Some research using particular species comes in for particular criticism, as might be expected. Dogs especially provoke concern: they are, after all, our "best friends." Individual researchers often draw the line at using such animals.

This is largely expressed as a personal choice, relying perhaps on experiences with dogs; yet, whatever personal choices a researcher makes about particular species, these are made in a context of wider stigmatization—not only about the fact of using animals at all, but also about using animals of those species. Doing painful things to dogs encounters more hostility in Western culture than doing painful things to rats or mice. Thus, a researcher who does decide to work with dogs (or who ends up doing so) is likely to experience greater stigmatization, and a greater need to justify his or her actions. As one technician recounted, in the course of an examination by his physician of many years, the physician asked, "So you are going to have to go back to work now and kill one of those cute dogs, huh?"

Several lab workers commented on the shift in public perception of scientific work using animals. Where once there had been greater, uncritical support for science, there is now much more overt questioning, on moral grounds. "I used to feel like a hero," noted one technician, "and now I feel like a criminal."[1] Sometimes, the feeling of stigmatization derives less from personal encounters and more from general signs of public criticism. One chief technician, for example, reported that everywhere he turned there seemed to be reminders that someone saw him as "bad." He explained that when he took his first lab job in the 1960s, animal research was seldom criticized publicly; he used to feel proud of his work and thought others regarded it highly. Now, there were anti-vivisection programs attacking animal research, and protesters demonstrating outside research facilities.

Although some research scientists also admitted being reticent if asked about their occupation, they are perhaps less likely than technicians to be put in that position because as physicians or academics they could talk about their work without mentioning animal experimentation explicitly. If they felt that the audience was not sympathetic, they might, out of frustration, tell people they "did cancer research" or "worked at Boston General Hospital." Only if they were convinced that listeners were reasonably sympathetic would they (gradually) reveal their work with animals.

One of the most significant metaphors—mentioned particularly by research scientists in the British study—was that of the fortress (also see Gluck and Kubacki, 1991). Shifts in public opinion due to the activities of animal rights groups had, they felt, left the scientific community "besieged." One spoke of how difficult it was to find the courage to "stick my head above the parapet" and speak publicly to defend animal use in research, while the scientist quoted at the beginning of this chapter acknowledged that peeping above parapets was precisely what was needed.

Part of being "behind the barricades" is that scientists sometimes feel that their community does not defend itself properly. One British researcher explained:

> I think [animal rights actions] have made scientists very frightened. It's also made them compromised because at a time when scientists were beginning to appreciate how important it was to try to explain and discuss their work, they were suddenly confronted, if they did so, with the potential of being firebombed or worse. It has in effect legitimized the squeeze on science . . . in the last few years. . . . I think the public unawareness and the inadequate defence of the scientific community of the use of animals has disadvantaged science in general. It has created a climate of opinion which is anti-science which has percolated through legislators and Government.

But, he went on to say, spokespeople for the scientific community did not always help the case if they were too outspoken. He avoided sticking his neck out too far because he was not sure he could do it well without fear of reprisal: "I am quite frankly not prepared to go to the barricades for it. Those who are prepared to go to the barricades for it should at least do it well. There's nothing worse than going to the barricades and losing."

Researchers seemed to adopt two strategies for dealing with the stigmatization and the feeling of being barricaded: they emphasized the need for public spokespeople (provided these were doing the job well) and they practiced what Goffman called "information control"—tactics such as evading the direct question, and speaking in euphemisms (similar to the tactics used in written papers, as we saw in chapter 3). But, several pointed out, controlling or avoiding information about work did not feel right; some felt an unreasonably imposed guilt. Referring to the comments made if animals were being moved through public spaces in cages, one technician said that sometimes "I will take a box down instead of the cage because you don't get so many questions if you're carrying a box. You learn to deal with it, but it's always there—that feeling that you're doing something wrong when you know you're not." That is, the very action taken to manage information could also make lab workers feel uneasy.

Managing Reproach

Most people involved in animal experiments are not self-stigmatized; that is, they see themselves as participating in a noble cause—finding cures for disease, alleviating environmental problems and so forth—and do not feel bad about it. When they do feel bad is when they perceive actual or threatened disapproval. Subsequently exercising control over information is, as Arluke (1991) points out, a common strategy people adopt when confronted with disconcerting social experiences. They must then try to make sense of what they have done both to themselves and to co-workers. They do so, he suggests, by carefully defining the "other" outside science as a threat, in one of four ways: that is, their speech addressed what might be called the reproaching, confrontive, dangerous, or distorting other.[2]

The use of "others" to establish a discursive space for a particular identity is not unique to those working in animal labs, we emphasize; it is well known in intergroup theory in social psychology, for example (Turner, 1985), in feminist critique of science (Halpin, 1989) and in the sociology of postmodernity (Lash and Urry, 1987). People come to define themselves and their identities, in short, by differentiating themselves from various others—it is a widespread habit that can serve to present one's own practices in a positive moral light. It may not, however, be enough to make people feel that those practices are properly accepted.

Some of the flak researchers reported came about, they felt, because of the behavior of some colleagues who did not "always do things properly," so giving the research community as a whole a bad name. In this sense, the "others" are inside science, but are seen as not upholding appropriate standards. One particular category was foreigners, who were often portrayed, in rather racist terms, as having lower standards regarding animal care. Asked, for example, about problems in relation to implementing legal controls over animal use, one British scientist felt that "I think perhaps where we have seen more of a problem has been foreign people coming in here, Orientals, Russians . . . where they've absolutely no regulations at all and it's quite often difficult to control these people." British scientists, moreover, sometimes referred to colleagues in the U.S. in the same way, suggesting that controls over animal use were less strong than in Britain, especially in relation to rats and mice.

A variety of other "others" outside science were identified in interviews as treating animals badly—the way that pets are often neglected by the public, for example, or the treatment of food animals in agriculture. The public would be surprised, we were often told, at how well the animals are treated in this lab; or the public is hypocritical in singling out science when ill-treatment of animals is so widespread. What is happening here is that the speakers are attempting to define a moral haven, a socioethical domain within which things are "done properly." For British scientists who were asked to reflect on the impact of legislation governing animal experiments, this helped to differentiate those insiders who were encouraged by what was perceived as a stringent law to "think carefully" about what they did, from those outsiders (such as foreigners) who were somehow less able culturally to engage in the painstaking ethical reflection required by the law.

It is, however, when dealing with outsiders to science that researchers are most likely to experience feelings of reproach. Many respondents said, for example, that they felt reproached by questions such as, "How could you do that?" a question which seemed to imply moral flaws. One technician reported being upset by a friend's mocking who jokingly referred to her as an "animal killer." Because the other person in these encounters appeared to be offended by the job, respondents felt criticized.

Encounters with strangers carries the added risk that it might become confrontational—precisely the risk that gives rise to the feeling of being besieged. Several said they were cautious about discussing their work "to avoid confrontation," and they often distinguished those who were "rational" from the "irrational"; that is, irrational others were ones who might become argumentative. "I like intellectual discussions," commented one researcher, "but you can't keep it intellectual. There is no exchange. It's just a ping-pong match."

Not surprisingly, several respondents referred to the hazards of admitting their work—the fears of reprisal by dangerous others. That fear means that people often seek to remain anonymous. "I don't want to be identified in this article by name," one biologist working in reproductive physiology in the U.K. asserted, "because there is an extreme fringe. You put not just yourself at risk, but your family. The extreme fringe is bombing cars and restaurants. This is a different kind of debate than we are used to even in the embryo lobby [referring to controversy over embryo research]. This is not rational debate."

While actual physical assault is relatively rare, it does happen. In one American hospital, for instance, a technician was kicked by a nurse as he was wheeling a dog onto an elevator (Arluke, 1991:321). Most information control to protect researchers from dangerous others takes place, however, through control over the physical setting—locked doors with card-coded entry, for example, or covering up low windows, as well as the location of animal houses apart from the main laboratories.

At times when rumors were circulating about particular labs being targeted by protesters, control of information was more likely. Experiments might be called off on a particular day if an anti-vivisectionist rally was nearby. Similarly, if unannounced inspections occurred, word traveled quickly; one senior technician, for example, traveled round to several labs to warn them, suggesting that "they make sure any animal you are using is under [anesthesia]."

Many researchers also feared distortion—being misrepresented by outsiders, particularly by the media. They then used information control strategies to prevent further distortion and to put animal research in a better light. One researcher told us that he had seen "how the press handle information which you give them, even the reputable press, it absolutely horrifies me. Once you've dealt with the press, you don't go into things with your eyes shut."

Given the possibility that researchers' comments might become distorted by the media, institutions are often careful about press releases. One American interest group for biomedical research, for instance, distributed information to coach laboratory personnel, suggesting that they emphasize that most lab animals are rodents and that pound dogs were not used in that state.

Potential distortion was a concern for many of our interviewees who asked how their words would be used. Arluke (1991) describes, for example, how researchers viewed him as a potential distorter when he first entered laboratories to do the research;[3] over time, he gained more trust. But, he points out, there is an irony in that pro-research advocates are themselves distorting: he experienced much pressure in his publications to eradicate words such as "stress" or "guilt," which could be used as "evidence" by the animal rights lobby. One reviewer advised "toning down" the piece to avoid making researchers sound bothered by what they did (even though that is exactly what researchers had often conveyed in interviews). In other words, they were suggesting distortion themselves to avoid subsequent distortion by others (similar to the history of editorial changes made in physiology journals, to which we referred in chapter 3; see Lederer, 1992).

In the face of public disquiet, and the threats of verbal or physical confrontation, researchers sometimes deploy tactics which, in effect, conceal information about what they do. Such tactics protect the individual (and his or her family and friends), and also help to build a supportive subculture; they do so, as we have seen, by disavowing their own moral flaws and pinpointing those of others. In this, they are not alone. People working in several occupations now find themselves the target of moral criticism based around the idea of a "helpless victim"—be that a forest, a fetus, or a laboratory ferret. For such people, finding strategies to manage information is crucial. It would seem, then, that the "general public" is regarded as harboring others who will, in one way or another, challenge animal researchers. The "general public" as a domain of casual encounters is thus a source of potential social risks that must be managed.

However, the general public is also what legitimates the work done by animal researchers—after all, biomedicine justifies its claims to authority (moral and epistemic) and funding because it claims to work for the benefit of "ordinary people." Moreover, in democratic societies, those institutions that call upon the public purse must, increasingly, address and incorporate the views of the public (see Nowotny et al., 2001; Irwin and Michael, 2003). Thus, the "general public" is a constituency to which pro-research scientists must explain themselves. Even if this explanation is disseminated by scientists' organizations and spokespersons, individual scientists still feel obliged to demonstrate that they are themselves willing to explain under the right circumstances. However, "explaining oneself" requires a public audience that will listen in the process of dialogue, which in turn means identifying components of that public who might be expected to listen.

Managing Rationality: Excluding Irrational Others

Managing information through concealment is, however, a strategy of self-protection aimed at dealing with those outsiders who, on the whole, are seen as presenting irrational arguments. These are seen as a threat not only to self and to one's peer group, but to science as a whole, or even the basis of post-Enlightenment culture. As one scientist explained,

> . . . there is a sort of strange way of, not just anti-science, but anti-intellectualist, a return [to] things which were unthought of, probably gone to mysticism . . . when an eclipse comes about, a return of anti-rationalism, it's actually quite worrying. And I think this is not restricted to science, it was to some extent a revolt against reason to a more primitive way of finding an answer to questions of life. It's very catching though, I mean it's, I don't take an apocalyptic view of this, but there is a trend that's running around.

However, researchers often distinguished such an irrational (and largely hostile) public from a more rational one, consisting of those who might be persuaded by careful argument. One way of doing that is to emphasize the regulatory system: rational people, that is, might be persuaded if they know there are tight controls. In Britain, where animal experimental procedures are controlled by statute, researchers often used this argument. So if confronted by reproach, they might point out that legislative controls are strictly enforced through the Home Office inspectors. Several emphasized that "their" inspectors were especially vigilant. The trouble is, we were told, that the public does not really know that there is a system of regulations, still less how it works (and, according to an opinion poll in the U.K., that seems to be the case; MORI, 2002).

In the specific context of scientists, however, differentiating others serves to demarcate the boundaries of what they perceive as rational behavior or beliefs. Interviewees emphasized the hypocrisy of the public with regard to pet-keeping or public demand for better medicine while simultaneously objecting to animal use in research. This claim typically took the form of pointing out how irrational and overemotional the public was, or how epistemologically naive about how science works. Attributing such flaws to outsiders or to the opposition is a typical move in any controversy—the anti-vivisectionists' arguments similarly hinge on arguing the irrationality and ethical dubiousness of animal researchers' positions, just as researchers do about their opponents. What it does here is to reinforce the belief that insiders—those participating in or supporting animal research—have privileged access to objective rationality. "Outsiders" are thus cast as overemotional and irrational.

At the same time, however, the scientists we interviewed often attempted to delineate a less irrational audience, consisting of "better-informed" members of the public. These people could, they implied, be won over if they were fully apprised of "the facts" about animal use and of the medical benefits that might accrue. The research community was, most felt, in danger of losing the battle because the opposition had achieved much stronger control over media messages.

As we noted in the last chapter, what is at stake for scientists is no less than a battle for hearts and minds: the public must be won over. In response to the growth and success of the animal rights movement, researchers have in recent years begun a counter-offensive, embracing a number of grassroots public relations organizations and patient groups. These pro-research groups have, in their members' own words, sought to provide public education and to "get the facts straight" about animal research in a "fight to get the mind of the public."

This campaign relies on what Best (1987) has called rhetorical resources—facts, justifications, and policy recommendations, articulated in symposia, conferences, and literature. But the resources used by pro-research organizations in these narratives do not just promote a product or technology, as scientists have done in other controversies (see, for example, Kleinman and Kloppenburg, 1991; Mulkay, 1993): what is at issue here is also the public image of biomedical researchers themselves, along with their current style of participation and debate.

Whatever individuals' motives might be, the implicit objectives of such public relations work are to seek a moral consensus. Through such public efforts, researchers are seeking jurisdiction over not only the scientific but also the ethical issues surrounding animal experimentation by giving themselves the moral authority to define acceptable and unacceptable research. This effort comes at a time when increasingly diverse publics—including scientists, federal authorities, animal welfare advocates, politicians, and others—have sought to have a voice in the debate.

The group addressed by scientists in this public relations exercise is primarily a rational one, people who could be won over by exposure to "the facts." Thus, one signatory to the British Association's Declaration on Animals in Medical Research in 1991 said he "strongly supported the Declaration . . . and hope[s] that the public will now be widely informed of the real facts" (Drury, 1991). We have seen how scientists sometimes express disquiet but are trained to write in ways that disguise what happens to animals in experiments. These facts, the details of animal use, are not ones they particularly want to disseminate; rather, what they seek to do is persuade the reasonable public of the merits of using animals in order to make medical advances (which signatories to the BA Declaration sought to emphasize), and to persuade them that "animals are well cared for." To do this, they must find rhetorical strategies that bring otherwise skeptical people into the fold.

Managing Membership: Whose Views Get to Count?

Appealing to rational audiences who might be persuaded to change their minds by appropriate arguments entails the attempt to create an in-group; this illustrates a rhetorical strategy deployed by members of what Harry Collins has called a "core set." Collins (1988) was studying the behavior of scientists engaged in scientific controversies, such as that over the relative safety of nuclear fuels. In the production of scientific knowledge, he argues, scientists must act to persuade colleagues of the value and authenticity of their work; in doing so, they also find ways to derogate opponents (through criticizing the quality of their work or deriding the status of their home institution). These two processes, of persuasion and vilification, act together to define a core set of people who are seen to be at the center of a particular scientific controversy and, thus, at the heart of the process of knowledge creation.[4] In other words, the core set is comprised of those scientists who, though they might disagree amongst themselves, nevertheless are fundamentally concerned with a specific controversial area of knowledge. Others—the wider public, or scientists who are not seen to belong to a pertinent discipline, for example—may be cast as outside the core set. In the case of the animal experimentation controversy, it is likely to be animal rights activists—seen as fundamentally irrational—who remain outside.

Yet in cases where scientific controversy clearly maps onto public concerns, as in the case of animal experiments, science must cast its net wider if it is to retain its authority. That is, scientists must enroll others to persuade them of the truth of the arguments. Partly this is because they need to justify the research in order to gain funding from public money. Enrolling others is, however, crucial to negotiating areas of moral controversy, such as using human stem cells for research or using animals. To continue such research means that at least some of the wider public (and its political representatives) must be convinced of its worth. Because, in principle, anyone—not only scientists—can have an opinion on issues of moral controversy, the core set can in theory be infinitely extendable. Less overtly, persuading more people is another tactic for dealing with stigmatization, by rendering at least some social encounters less dangerous.

However, widening the net also means, inevitably, that defining core sets becomes rather messy. In our interviews, scientists generally sought to demarcate a core set—that is, to define those people, inside or outside of laboratory practice, who might have proper membership, whose opinion might be worth noting (Michael and Birke, 1994). This is not a question of right or wrong, but a question of establishing who might be disqualified from commenting or judging; as we have seen, our interviewees tended to categorize certain "others," such as foreigners, as outsiders on the grounds that they were not adhering to particular moral codes.

Collins documented the rhetorical strategies used by scientists to undermine their opponents within the core set. Other groups of professionals might, of course, use similar strategies of persuasion, but in the case of science, what is at issue is the "truth" about a state of affairs in nature. Winning the arguments is thus about establishing truth claims (see Latour, 1983). If scientists cast doubt on the quality of competitors' science (or doubt the competitors' personal integrity), then the "facts" yielded by their experiments will also be doubted.

The case of animal experimentation is slightly different, however. Here, it is not so much a case of persuading rival groups of scientists on points of technical merit, but rather a case of persuading wider publics using an overtly moral component: whether it is right or ethically defensible to use animals in pursuit of scientific knowledge or medical benefit. This makes it rather different from other controversies, where a key feature is an emphasis on a core set of experts, usually those with particular technical knowledge. Although Collins argues that controversies might be opened up to wider public debate (Collins, 1988), his view has been that this happens through representative experts—in the case of the debates about the safety of nuclear fuels, this might, for example, be organizations like Greenpeace. This suggestion makes apparent sense, because the core set in this case is characterized by technical expertise, though, as numerous critics have argued, what counts as relevant expertise is itself controversial and extends beyond a strictly defined scientific core set (see Collins and Evans, 2002; Wynne, 2003).

Be that as it may, if we draw on the perspective of actor-network theory, we see an important set of issues emerge. According to "classical" actor network theory (ANT), put rather simply, scientists must have certain practical, political, social, and economic skills—that is, their powers of persuasion are seen to reside in much more than just technical skill in the lab, and these matter just as much as technical details in understanding scientific controversies.[5] Thus, scientists marshall a range of materials and techniques to extend their influence beyond the laboratory. To do this, they must enroll others. One scientific actor can use various strategies to enroll other entities to help them make their case, such as other scientists, publics, animals, or even electrons. In the case of animal laboratories, scientists have over time enrolled particular kinds of animals into the practices of science, along with a range of physical apparatuses and external institutions that support that kind of animal in that kind of experiment. They can also be said to be enrolling modern lab animals when scientist-advocates emphasize that "nearly all the animals used in research are purpose-bred." That is, they construct a story about purpose-bred animals in an effort to marshall the argument regarding the acceptability of using animals.

Yet the technical is only part of the enrollment. In Collins's (1988) study of a public demonstration of the safety of nuclear waste flasks, the public was asked to take on a range of identities in the controversy—temporary experimenter (in observing a mock experiment), citizen concerned with national economic well-being, and so on. In doing so, they were asked to dissociate themselves from other identities (such as rejectors of nuclear fuel use). Thus in this particular controversy, scientists speaking on behalf of nuclear fuels were more than simply experts; they also acted to enroll others through political and economic arguments.

When these broader political, ethical, and economic issues are also incorporated into a controversy, the core set becomes much wider. That is, the core set now comprises all those actors having a concern with moral and political aspects of the controversy. This could, theoretically, encompass any member of society with a view on the relevant subject. Since the animal experimentation controversy is overtly moral in character, the allied core set is, potentially at least, infinitely extended—any person can have a moral stance on this issue.

Scientists, however, must attempt to constrict the core set, not least by excluding those whose views are deemed irrational. Scientists involved in animal experimentation, finding themselves embroiled in a messy and disparate core set spanning many issues, perspectives, interests, and even cosmologies, attempt to restrict and redefine it. As such, scientists are seeking to demarcate the boundaries of the core set: they want to stipulate the terms of the debate. Accordingly, on the whole, it needs to have certain minimal characteristics—rationality, nonviolence, civility, and so on.

The more amenable—more rational—section of what might be called the "concerned public" was consistently invoked in our interviews. This section included moderate animal welfarists (such as many members of animal protection organizations), and (some of) the general public, with whom debate is already possible (e.g., patients and their lay allies) or will eventually become possible. Yet, although scientists attribute rationality to the moderate faction, they simultaneously show that rationality to be compromised. Thus, while pro-research narratives usually did not demonize the public or flatly rule out its participation, its voice was still problematized. That is, unless members of the public could be taught to think and feel like scientists—tempering their emotion and becoming more informed—the public should also be disqualified from participation in debate. In other words, the public's rationality was judged not simply in terms of form (a reasonable style of argument), but also in relation to content (specific expert knowledge, not least about animals).

Consistently, pro-research spokespeople questioned the capacity of even the moderate public to debate seriously by portraying it as ignorant about science or animals, thus rendering it ill-equipped to advise on any policy related to animals in science. At a seminar of the American Medical Association, for example, one speaker lamented an anti-science attitude: "ignorance is the problem—no knowledge of the scientific method." Members of this anti-science minority, in his opinion, were "contemporary luddites." He went on to cite a 1957 study by Margaret Mead of 35,000 high school students' stereotypes of scientists—one element was that students saw them as cutting up animals and injecting weird serums. Others cite research documenting an apparent lack of public knowledge or understanding of science, including one survey suggesting that nearly all American school children are scientifically illiterate—"they don't know which comes first, thunder or lightning," observed one commentator.

Pro-research spokespeople sometimes identify several aspects of ignorance thought to disqualify the public as a partner in debate. Being uninformed, the public would be too confused, amateurish, or irrational to address the issues seriously and to weigh the costs of benefits of research. One speaker at a conference noted, for example, that because so few people now lived on farms, many held a "romanticized notion of animals based on our experiences with pets." A limited experience with animals, in short, meant that people's understanding of them would be distorted, leading to inappropriate or wasteful suggestions. Researchers at a workshop on primate well-being, for instance, lamented lay suggestions to increase cage size because they were based on anthropomorphism rather than scientific understanding of the psychological needs of primates.

Most interviewees felt that the public did not understand the significance of animal research in achieving medical advancement, which should disqualify such people from sitting on regulatory committees. The chair of one such committee told us: "The average person doesn't have all the information or does not spend the time to think about the implications of animal-based research, the implications of the loss of animal-based research, and therefore society's views are perhaps somewhat less relevant than our own in that respect."

As with the accusations levied at animal rights activists, some research advocates attributed this public ignorance to popular preoccupation with "pseudo-science," such as astrology columns in newspapers. One complained that Americans were more interested in sex and entertainment on commercial television—"the idiot box"—than the educational documentaries public television stations had to offer. Who watches that, one wondered?—"Nobody. It's not sexy. It's not interesting. It's not entertaining. It's not fun."[6]

Thinking Harder, Caring Harder: Who Has the Expertise about Animals?

What is going on here is that the spokespersons for the research community are trying to identify who has sufficient expertise to enter the debate. The irrational part of the public (those who watch entertainment) does not. At the same time, researchers appeal to those more serious members of the public who might be willing to listen to "the real facts." In doing so, they seek to establish the legitimacy of their own voice by portraying themselves as moderate centrists who care for animals and who might therefore engage the concerned public in debate.

Yet researchers simultaneously distinguished themselves from the public in ways that could give them greater voice: in particular, they often claimed to possess expertise needed to judge properly the use of animals in research—expertise from which the lay public is excluded. Some scientists argue, for example, that they possess superior knowledge derived from their training in ethology, which privileges their perception of animal pain and welfare. Technicians similarly referred often to their own knowledge of animals contrasted to the "ignorance" of much of the public. In short, researchers were attempting to define and delimit a core set of those who might have sufficient expertise to evaluate experiments.

Sometimes this "expertise" meant intellectual or technological skill; at other times, it referred to practitioners' belief in their profounder knowledge. By contrast, the alleged greater emotionality of the lay public would disable proper reflection. Speaking about the change in British law after 1986, several scientists spoke of how the new legislation had usefully focused their thoughts, forcing them to think about what they had been doing. This "thinking harder" serves as another criterion for membership of the core set, since researchers often questioned the capacity or willingness of the lay public to think hard: rather, "they [the lay public] have this vision that [scientists] are all slightly mad and you are all waiting for the opportunity behind closed doors to inflict pain and misery on animals. It's terrifying that anyone could think that," one researcher commented.

What this quote implies is that the public's response is basically irrational, perceiving scientists in simplistic terms as evil and intent on pursuing benefits at any cost. That is, the public is represented as resorting to stereotypes rather than engaging in the "hard thought" characterizing the research community's efforts.[7] The necessity for hard thinking as a criterion for admission to the core set is underlined by some scientists' suggestion that ethics professionals might be admitted—that is, people who have expertise in the calm, considered calculation of the value of scientific experiments. For example, in the following extract, a scientist remarks on the role and composition of ethics committees that might adjudicate on the necessity of experiments:

> Thankfully what goes on in ethical committees in general is reasoned debate. All the ones I've had contact with have worked very effectively. For example, most lay representatives would be members of the clergy or legal profession—in other words they would be people who, as part of their daily lives, were used to judging ethical situations. There are even people who are professional ethicists. That is not inappropriate—judging the ethical position on any given subject is quite a professional activity. You can't just walk in and give a gut reaction to what you are being told. You have to carefully consider the pros and cons. There is a tradition of ethics you really have to be aware of. It is important that lay people are selected so that rational debate can take place.

Here, the speaker alludes to a stratum of ethics professionals who can argue the issues in accordance with a rationality more or less absent from the amateur lay public. But it is not emotionality per se that so disables the layperson from rational debate in these portrayals; rather, it is an emotionality that reacts to superficial appearances, rather than on the basis of appropriate experience of animals or understanding of the possible benefits.

Yet scientists cannot appeal to the wider public if they dismiss emotions altogether. Indeed, they are often encouraged to show feelings for lab animals or pets, which is seen as an appropriate expression of emotionality. A speaker at one seminar in the United States acknowledged that researchers might be hesitant to do this: "I know it sounds heterodox to say, but we should express our concern. Some of you will think this is getting into bed with the enemy, but mostly people want to know that you care for animals—so say you have pets and care about animals."

Scientists often spoke in interviews about their affection for their own pets, as well as for lab animals, as we have already seen. At symposia and public hearings, researchers declared animals "partners in research" (as they are portrayed in the advertisements we discussed in chapter 3). One American pediatrician told his audience: "I close with a word of thanks to those dogs who provided what was necessary for my 89-year-old mother to have both her hip joints replaced." Others refer to benefits to animals in the form of protection against rabies, canine distemper, or parvovirus. One Nobel laureate thus spoke at an American Medical Association meeting about his research on kidney transplantation, saying: "I love animals. I have a dog and a cat a home, my grandchildren also do." He went on to talk about the dogs he used for his transplantation research, showing a slide of one of these dogs seated behind its birthday cake: "They became pets for all of us. Many were taken home as family pets," he added. Those who attended the workshop were given, in their resource kits, a prepared talk with accompanying slides, one of which (#36: AMA, 1992) reiterated: "Scientists aren't white-coated frankensteins. They are literally the people next door. They have families and pets. They have sons who love their dogs and 12-year-old daughters who are crazy about horses. Their compassion is the first line of defense against cruelty."

While the research community may thus appeal to the emotional responses of its members and their liking for animals, the emotionality of laypeople is often rendered dubious or suspect. As one researcher commented, echoing the fear of anti-rationalism noted by others, "I think there is a zeitgeist feeling of concern for animals . . . and I think a lot of those people [who] get sucked into that are just people who are angry about how they are brought up." In other words, public emotional responses are inauthentic or illogical—

particularly in the apparently hypocritical way they castigate experimenters while using the products of medical research or mistreating their own pets. A typical feeling from researchers was this: "I have seen what the public subject their pets to, which is unbelievable." What is being emphasized here is the superficiality of the public's emotional response to animals: it is a sentimentality that has, we were told, attached itself to lab animals while neglecting other cases of cruelty to animals nearer home. Yet again, public emotionality is represented as inferior.

Confronting the Public

In this chapter, we have illustrated how, on a personal level, laboratory workers sometimes experience stigmatization as a result of public hostility. Technicians, particularly, talk about stigma, and how they must find ways of rhetorically dealing with possible reproach or attack. Scientists, too, report feeling besieged and under threat. But to a much greater extent than technicians, they can mobilize and produce counter-arguments based on the quality of the science. Indeed, there has been a very clear campaign in recent years to oppose publicly the claims of animal rights organizations and to present the researchers' case. What this relies on, we have argued, is attempts by scientists to delineate and specify those outsiders whose reasonable judgment might be admitted. They must also tackle the tricky terrain of emotions, acknowledging some of their own emotionality as, for example, pet-owners, while diminishing the (over)emotionality of sectors of the public. So for all the emphasis on rationality,[8] emotions—passions—do run high throughout, and on both sides of, this debate.

In these appeals to a wider audience, researchers are establishing particular identities or roles, defining those people outside science with whom there can be engagement. These identities are characterized by formal qualities: they concern the basic conditions for controversy and debate to be conducted in what is, from the scientists' perspective, an appropriate and orderly fashion. Thus, while the form of argumentation should be logical, authentic, hard-thinking, emotionally modulated, and so on, the content of the arguments is left open. It is, however, assumed that the relations or associations between anti- and pro-animal experimentation actors will be antagonistic; the expressed desire is that the antagonism should take a "rational" form. At the same time, as we have seen, such formal dimensions of argumentation are compromised by what are seen as low levels of factual knowledge. Thus, public suggestions based on perceived ignorance, however rationally articulated or civilly offered, are regarded as irrational in the context of what scientists take to be "the facts" (a complaint similarly made in wider contexts regarding public understanding of science, as we discuss in the next chapter).

In this chapter, we have traced a number of ways in which pro-research scientists represent the "general public." On one hand, the public is a source of potential criticism and reproach, which can lead to a sense of stigmatization. In this sense, the public is not something to be confronted directly; rather, researchers find themselves taking evasive action. On the other hand, the public is also something to be confronted. This may be through specific spokespeople, brave enough to "come out of the closet" and present the case for research. More often, the confrontation is indirect, through researchers' attempts to define lay outsiders as ignorant and lacking expertise.

These ways of representing the lay public do not arise informally simply within pro-research culture, however. The general public is an object of social scientific study—its views, attitudes, opinions, beliefs, and levels of scientific literacy have all been subjected to scrutiny. It is hardly controversial to suggest that the results of such scrutiny are one resource in pro-research scientists' own beliefs about the nature of the "general public." Yet such scrutiny is never innocent: it inevitably entails assumptions about what we mean by "the public" (and its relation to scientific institutions). In the next chapter, we summarize some of the findings about the general public's views on animal experimentation, but also unravel some of the problems with these findings, not least as they arise in the complex context in which the scientific uses of animals, the symbolic standing of animals, and the relations between publics and governments seem to be undergoing substantial change.

IX

Making Publics, Scientists and Laboratory Animals

In Douglas Adams's *Hitchhiker's Guide to the Universe,* white lab mice, rather than being the mere objects of experimentation, turn out to be multidimensional beings who ironically experiment upon their scientists. The joke here is that scientists, for all their brilliance, have very little idea about the nature of the world beyond their immediate objects of study. Nowadays, there is an additional irony to be mined in this image: to work with laboratory animals is to be part of a social experiment in which scientists find themselves the subjects.

Similarly, Warner Brothers' television cartoon series about two lab mice, *Pinky and the Brain*, ironizes science as both brilliant and stupid. In episode after episode, Brain, a mouse that happens to be a scientific genius, embarks on yet another inspired attempt at world domination only to be thwarted by his idiotic companion Pinky. The joke is that Brain seems powerless to plan for the unpredictability of Pinky. In this case, overconfident scientists—embodied ironically by Brain—are routinely brought low by the (predictable) unpredictability of the world (Pinky).

In both these representations, lab animals—and scientists—have complex meanings. As we have seen throughout this book, lab animals are culturally highly important in many ways. They ambiguously signify both the selfless biomedical enterprise, which has putatively yielded so many benefits, and the perceived hubris and arrogance of scientific institutions and scientists. This ambiguity is, of course, deeply rooted in Western science—we have seen in the last few chapters how it emerges in the way that scientists express themselves. Ambiguity is also routinely expressed and explored in popular culture, where

173

laboratory animals are a complex cultural sign, mediating various anxieties, concerns, dilemmas, and identities—as the two examples above illustrate. And these complexities are exacerbated further in the face of a changing relationship to nature, wrought by new developments in biotechnology, that influences public attitudes.

Yet if both animals and scientists have multiple meanings, what about "the public"? While the wider public is engaged with the complex cultural meanings of animals, scientists, as we have seen, routinely draw on an image of the general public as ignorant or irrational in relation to the politics of the animal experimentation controversy. So "the public" is both embroiled in the production of multiple meanings of animals (and lab animals in particular), yet at the same time is prone to ignorance, emotionality, irrationality, and even violence. Complicating matters still further are the ambiguous findings from surveys of public opinion on various aspects of science (such as animal experimentation) or, indeed, science in general.

By and large, studies of how the public views animal experimentation have been geared toward surveying attitudes or levels of knowledge rather than exploring broader cultural dynamics. Surveys of attitudes and understandings are, however, politically important for continued support of science from public funds. This has been a major reason for the extensive surveys of the public's knowledge of, and attitudes toward, science carried out in recent years, particularly in the U.S. and Europe (Wynne, 1995). Underlying much of this research is the view that negative attitudes toward science are the result of poor scientific literacy, or simply ignorance of what are seen to be the scientific facts: if people knew more, the argument runs, they would be more positively disposed toward science (e.g., Irwin, 1995). As we have seen, this is a familiar argument in the animal experimentation controversy.

Yet surveys have come in for criticism, for two reasons: first, there is a problem in assuming an equation between knowledge and support, as several studies of public understanding have shown. On the contrary, publics, much like scientists themselves, believe the bodies of knowledge of those toward whom they are positively disposed. Facts, that is, rest on trust: if one trusts a scientist or politician, one is more likely to assume that their statements of fact are true or credible (Wynne, 1996; Michael and Carter, 2001; Irwin and Michael, 2003).

Second, surveys do not simply record a single "public." We have seen, for instance, how various publics can be differentiated in terms of their knowledgeability, rationality, and so on, regarding animal experimentation. Thus, the public is not a simple entity that exists for all to see. It is something that, as several authors have argued, is "made"—its qualities, its characteristics, its borders, have all changed over time (e.g., Shapin, 1991). Surveys, moreover, help to generate an image that shapes what it means to be a member of the public, and how to comport oneself as a member of the public (e.g., Hacking, 1986). For example, while we now talk about having attitudes and opinions, these are relatively recent inventions, and various institutions have had to work hard to make such constructs not only acceptable, but part of our psychological vocabulary (Osborne and Rose, 1999).

Given these criticisms, it is important not only to explore survey results, but also to trace the assumptions that lie behind such analyses, assumptions that influence what comes to be seen as a "good" or "bad" public, and indeed good or bad scientific institutions. We can also draw on qualitative studies (for example, semistructured interviews and focus groups) that complement survey approaches and grapple with the complexity of people's views on animal experimentation. These tend to have a specific focus (people's attitudes toward nuclear fuels, for instance) and to allow study of people's trust in institutions. Here, a different model of the public (and indeed, of "understanding") operates—one that places emphasis on trust rather than understanding, and on identity rather than knowledge.

Alongside these studies, there is a growing movement to democratize scientific governance—that is, to develop methods enabling the public to have a voice in scientific policy and decision-making.[1] The implications of this are, potentially at least, enormous, not least in terms of discriminating the appropriate publics—scientific citizens—that can be invited to contribute. As we saw in chapters 7 and 8, pro-animal research scientists have very particular views about what sorts of lay people should be permitted into the process of evaluating animal experiments. Possession of such qualities as "rationality" and the right sort of "emotionality," where emotional expression is subordinated to what are perceived as rational principles of debate and argumentation, is seen to be a prerequisite for a place at the conference table.

Here, then, we see very different versions of "the public" being deployed across these quantitative, qualitative, and participative methodological approaches. Respectively, these are bearers of knowledge and opinions often in need of education; social beings whose identities influence whether they trust this or that source of knowledge; and "scientific citizens" whose participation is important in democratic scientific decision-making. These different versions of the public relate to different versions of both science and scientific governance, and, by implication, the symbolic role of laboratory animals, as we shall see.

Surveying the Surveys: What Does the Public Think?

Finding out what people think about animal experiments by analyzing surveys is harder than it seems. In the first place, it would be impossible to review surveys exhaustively, not least because only some are fully published, being commissioned either by the media itself or by advocacy groups (Herzog, Rowan and Kossow, 2001). Moreover, the value of some surveys is somewhat compromised when, as these authors comment (and this is a hazard that extends well beyond the issue of animal experimentation), the questions that make up these surveys are worded in highly biased ways.

Given that, we will not present a comprehensive survey of the surveys, but will examine public attitudes toward laboratory animals and animal experimentation in terms of key parameters. That way, at least we would get a sense of some of the factors that are more or less reliably associated with views on animal experimentation. These parameters include demographic variables such as age, gender, and political affiliation but also factors pertinent to the experimental use of animals (e.g., species, procedure, pain and suffering, analgesics, medical benefit, alternatives).

Over time, at least since the 1940s, there seems to have been a general movement toward a less positive attitude to animal experiments (see Herzog, Rowan and Kossow, 2001). In terms of demographic variables, women and younger persons tend to be less favorably disposed to animal experimentation than men and older persons (e.g., Medical Research Council, 1999). Indeed, among young adults in the U.S., "taking into account gender differences in feminist attitudes, attitudes about science, scientific literacy and early encouragement in science, women were still more likely than men to oppose animal research" (Pifer, 1996, p. 10). Political affiliation or leaning is also correlated with attitudes. Generally speaking, the more right-wing, the more positive the attitude (Eurobarometer 55.2, 2001; Herzog et al., 2001).

One consistent theme in surveys is that people often draw distinctions between different types of animal use in laboratories (just as many scientists do, as we noted previously). The most obvious distinction is between laboratory animals used for medical experiments, and those used for cosmetics testing. It is very common indeed that the former are more positively regarded than the latter. For example, Aldhous, Coghlan and Copley (1999) found that whether or not mice were subjected to pain, illness, or surgery, publics were (up to four times) more likely to disapprove if the experiment was designed to test the safety of a cosmetics ingredient than if it tested the safety and effectiveness of a drug or vaccine.

Unsurprisingly perhaps, surveys also indicate that people differentiate between species: they are generally more likely to find experimentation acceptable if it is conducted on mice and rats. For example, Herzog et al. (2001) review studies in which the proportion of the public expressing concern for the general welfare of animals decreases as we proceed down what amounts to a hierarchy of species. Thus, 89% of the sample express concern regarding dogs' general welfare, while 34% express concern for mice. Reflecting this difference in the specific case of animal experiments, 51.3% support experimental use of dogs, whereas 76.1% support the experimental use of mice (and 59.5% for monkeys). The difference in the support (or opposition) to the experimental use of specific species does not depend on whether the animals involved are subjected to pain or not, although overall support is reduced when animals are subjected to pain. That is, another key factor influencing people's support for animal experimentation is whether experimental procedures feature pain or suffering (Aldhous et al., 1999; MRC, 1999).

Other factors inevitably complicate this picture, such as the perceived availability of alternatives to animals; the severity of the medical condition; the perceived standing of the groups likely to benefit from the experiment (e.g., children versus gay adults); the stringency of the regulatory regimes overseeing animal experimentation, and the sense that animal welfare was properly attended to (MRC, 1999; Herzog et al., 2001; Breakwell, 2002).

Some studies (e.g., Pifer, 1996) suggest that scientific literacy was not a major factor in determining attitude. By contrast, the MRC (1999) survey in the U.K. found that "many of the groups who are more knowledgeable are also more supportive of animal experimentation" (p. 32). The temptation here is to infer that more knowledge will yield more support. Certainly, as we noted above, a similar assumption has colored the views of scientific institutions seeking support in the past (e.g., Royal Society, 1985; Wynne, 1995). But as the MRC report goes on to suggest, there are several ways to explain this correlation. For example, men, who are generally more positively disposed to animal experimentation, are also likely to be more knowledgeable. That is to say, as with Pifer's study, support might have more to do with gender than knowledgeability per se.

"Knowledgeability," moreover, is not a simple construct. It can refer not only to "facts" but also to the trustworthiness of the source of that knowledge. Thus, in the MRC survey laypeople were asked to judge the "percentage of medical research that involves animal experimentation," yet "percentage" might mean several things: percentage of research funds allocated; percentage of all medical research projects; or percentage in terms of numbers of human and animal subjects. There is, it would seem, no simple "fact" about "percentage of medical research that involves animal experimentation." Nor is the meaning of "medical" research obvious or uncontroversial: it can mean all research funded by the Medical Research Council or only that research having immediate health benefits.

This problem of meaning is particularly acute as various studies have shown that people are more likely to disapprove of basic experiments that, for example, "study how the sense of hearing works" (Aldhous, Coghlan and Copley, 1999) or "learn how cells work" (MRC, 1999), than studies that are directed toward specifiable treatments. While, quite understandably, scientific institutions see such research programs as part of medical research, this is not necessarily the case for people outside them.

To refer to "percentage of medical research that involves animal experimentation" thus involves several assumptions that might stand at odds with assumptions made by lay respondents. The danger here is that scientific institutions, in trying to promote greater public knowledge, end up alienating publics by virtue of presupposing what counts as "medical research" and "percentage." Laypeople are not unaware of these slippages in meaning, so part of being "scientifically literate" or knowledgeable might have as much to do with a sense of such slippages and the uses to which they can be put (that is, the political context of science) as understanding particular facts (Durant, 1993).

Scientific literacy might be better conceived in terms of an awareness of how scientific institutions work and how laboratory work proceeds in practice. If people are already predisposed to regard scientific institutions with suspicion in contemporary social and political circumstances (Giddens, 1991; Beck, 1992), then efforts at raising scientific literacy about animal experimentation are likely to be seen as examples of what Wynne (1991, 1992) has called "institutional body language." Such body language, expressed most obviously in scientific spokespersons' highly confident statements of fact, serves to obscure the complexities and uncertainties of scientific process and products (knowledge). As a result, scientific institutions can seem to be imposing what is to count as knowledge, so that public suspicion and skepticism will actually be reinforced. We shall return to this point below.

From this brief overview a picture emerges in which people's support of animal experimentation is affected by demographic variables (such as age, gender, politics), as well as aspects of the research process itself (type of species, levels of pain, perceived relevance to medical benefits). The typical survey approach has not, however, been particularly adept at accessing the ways in which these various factors work together—that is, how people actively weigh these considerations of species, utility, or pain.

Calculating Costs and Benefits

How, then, do people weigh the issues? One survey (Aldhous et al., 1999) attempted to explore just this. It began by asking people about animal experiments in two ways. To begin with, respondents were asked if they agreed or disagreed that scientists should be permitted to conduct experiments on animals; they were then asked again, after an explanation that scientists were developing drugs to reduce pain and treatments for two different life-threatening diseases: AIDS and leukemia. In the first condition, a mere 24% were in favor of animal experimentation, with 64% against. But in the second condition, these proportions changed dramatically: 45% were in favor, 41% against. The swing for both men and women was around 20%. Those whose views swung the least were, not unexpectedly, members of animal welfare organizations.

The argument here is that if laypeople are made aware of particular facts (here, the reasons behind animal experiments), they will weigh the pros and cons. This was reinforced by the finding that such reasoning also took into account species of experimental animal, levels of pain, and the objectives informing the experiment (from curing childhood leukemia to cosmetics testing). As the heading of one of the tables in the report stated: "People carefully weigh the costs and benefits of individual experiments before deciding whether they approve."

Indeed, the report gives the impression that the public's capacity to engage in cost-benefit thinking is something to be celebrated. Writing about the attitudes of members of the Japanese public toward using animals for xenotransplantation,[2] Mercer et al. (2002) even seem to suggest that cost-benefit thinking should be seen to be cross-cultural and a sign of "bioethical maturity." They write: "If we consider bioethical maturity as a ratio of those who consider both benefits and risks (specifically of xenotransplantation), then the [Japanese] public could be argued to be mature in this sense" (p. 359). Once again, we see a highly rationalistic model of the lay public being highlighted, and potentially emotional responses being downplayed. The "ideal" member of the public should have certain cognitive qualities, namely those that can consider both risks and benefits, pros and cons. However, this model of the "good" or mature public entails a series of problematic assumptions, not least about the viability of cost-benefit thinking. Making these kinds of judgments is by no means as straightforward or as rational as many surveys seem to imply.

In fields as varied as environmental impact assessment, health policy, and technology assessment, critics (e.g., Foster, 1997) have pointed to numerous problems of technical cost-benefit analysis: that costs and benefits can proliferate; the identification of costs and benefits is chronically negotiable; costs and benefits entail tacit cultural assumptions and incorporate implicit relations of trust. Indeed, many of these concerns are articulated by scientists themselves, who, in the U.K., are required explicitly to justify their experiments to the Home Office in terms of costs and benefits. As Michael and Birke (1995) illustrate, despite being readily able to justify their work in cost-benefit terms, scientists also reflected on ways in which these were uncertain—such as not always knowing just how much pain animals suffered, or not being able to predict confidently potential medical benefits. Even for scientists, then, cost-benefit analysis cannot be done wholly at a rational level.

The limitations of focusing only on rational decision-making in ethical thinking is clear in Michael and Brown's (2003b; 2004) focus group study of laypeople's views on xenotransplantation—a topic posing many ethical dilemmas about human health and animal use. Here, there was certainly evidence of cost-benefit thinking, often as a collective endeavor in which participants "cycled through" the pros and cons, articulating various dilemmas about xenotransplantation. A typical cycle might include the statement that xenotransplantation is another example of scientists "messing with nature," that is, it is unnatural. This would be followed by the view that despite this, one would always opt for xenotransplantation if it was the only means to survive. This might be contrasted to an expression of concern for animal welfare, which in turn would be countered by the observation that animal bodies are routinely used in many areas of medicine (e.g., pig heart valves) and for consumption (meat). This cycling through of pro and con arguments rarely, however, resulted in resolution or consensus. Moreover, the process of argument—this cost-benefit thinking—was *itself* tacitly scrutinized by the participants.

Michael and Brown identified three discourses (what they called meta-arguments of *trust, telos* and *trump*) by which laypeople showed that they were not entirely persuaded by their own processes of reasoning. These can be summarized as follows:

> *Trust:* Laypeople were deeply aware of having to make judgments of trust in order to know what is being compared with what in the process of cost-benefit thinking. For example, they might draw contrasts between the sensationalism of animal welfare groups versus the sober objectivity of governmental bodies, or the investigative "willingness to expose" of the animal welfare groups versus the bureaucratic "tendency to secrecy" of governmental bodies.[3] That is, both animal welfare organizations and governmental bodies can be both trusted and distrusted. This pattern

throws into relief just how contingent such trust, and thereby the related process of cost-benefit thinking, can be. Here, it is possible to identify a "meta-responsibility" on the part of the participants: their citizenly activity takes note not only of the pros and cons, but also of the contingency of the elements that go into a cost-benefit calculation.

Telos: Laypeople often commented that *their* cost-benefit thoughts on xenotransplantation are irrelevant to any broader process of decision-making. Such innovations are, they felt, inevitable given the way science operates (the various economic or social factors which drive biomedical innovators and entrepreneurs—preeminently the twin desires for profit and fame). This seems to reflect a different citizenly attitude—on the one hand it is an attitude of withdrawal from a particular form of calculus, and on the other it implies a rather different type of citizenly activity (something which is perhaps more radically politically interventionist).

Trump: This concerns the perception that whatever concerns or ethical position one takes, these will depend on whether one is facing death or not. This is, in effect, a "bottom line" argument: when push comes to shove, one will try anything to survive, whatever the consequences. Here is a vision of a generic life—life per se—which trumps all negative valuations of relevant biomedical innovation. But this argument was itself further nuanced by issues of quality of life. That is to say, participants would note that certain lives might not be worth living. In a different context, there is also a process of what might be called "reverse-trump." Thus Irwin and Michael (2003) have noted that certain laypeople, despite being very ill, will argue against animal experimentation even where this might lead to eventual benefit for them. In this case, cost-benefit thinking is undermined by deep moral convictions rather than a more "primitive" will to survive. What this means is that people use alternative rationalities in their deliberation: pros and cons and their calculation are overtaken by values such as desire for dignity (to die with dignity), respect for others (not to burden carers), and a commitment to animal rights.

These, then, are some of the ways in which people talk about the biomedical use of animals. To be sure, we have concentrated on xenotransplantation—a somewhat peculiar and, thus far, barely realized use of animals. But this suggests important insights for the ways in which laypeople deliberate over animal experimentation in general. Trust, telos, and trump are discourses likely to have widespread currency. Laypeople are more than calculators of pros and cons—they evaluate such controversial uses of animals in ways which draw upon a much broader sense of the cultural, social, economic, and political context in which such controversies arise and circulate. Indeed, such an "expanded" rationality that locates the calculation in complex contexts might well be seen as irrational from the perspective of scientific institutions (which, perhaps necessarily, operate with a more restricted cost-benefit rationality). This throws a rather different light on researchers' charges of public irrationality that we explored in chapters 7 and 8.

Yet, although animal experimentation in general and newer techniques such as xenotransplantation may elicit similar reactions from people, there are also differences. In particular, new techniques seem to evoke a whole new range of public concerns about animals, especially if those techniques involve genetic manipulation. What is at stake here is a rapidly changing relationship to nature.

From Strain and Model to Hybrid and Product

In the early chapters of this book, we explored ways in which laboratory animals developed, and the uses to which they have been put. On the whole, we considered these historically, focusing primarily on traditional methods of breeding and selection. Similarly, in examining the laboratory animal as an experimental subject we broadly focused on its role as a "model" which stands in for human bodies. This traditional figure of the selectively bred animal serving as a model is, however, rapidly being transformed—in particular by the widespread use of new genetic technologies to transfer genes from one species into another. Such newfound capacity to intervene radically and speedily in the composition of a species' genome has, according to some commentators, fundamentally changed our relationship with nature; we have thereby entered an epoch of what Paul Rabinow (1996) calls *biosociality*. In this, "nature is remodelled on culture understood as a practice. Nature will be known and remade through technique and will finally become artificial, just as culture becomes natural" (1996:99). Accordingly, the biological is both a product of human invention (with all the cultural assumptions that inform such invention), while at the same time, humans are increasingly reduced to biologistic, genetic entities.

If we are indeed entering the epoch of biosociality, it is not one in which laypeople are necessarily comfortable. Looking at British laypeople's views on animals and biotechnology, Phil Macnaghten (2001) found that there was considerable concern about the genetic manipulation of animals and the uses to which such transgenic animals might be put. He asked laypeople to consider a number of possible future applications for genetically modified animals including sheep and cows that make pharmaceutical products in their milk, faster-growing farm animals and fish, and pigs that are bred to produce organs that can be transplanted into humans.

Perhaps not surprisingly, the responses were overwhelmingly negative. Partly, this was a pragmatic response, as people questioned the seeming usefulness and possible risks of such biotechnological innovations (Macnaghten, 2001). These two dimensions often combined as concerns about violating natural boundaries and the capacity of nature to fight back with a vengeance, or that "messing about with nature" for the sake of profit or hubris was "likely to rebound on humans" (the BSE or "mad cow disease" crisis in the U.K. was a touchstone in this regard).

Participants were also explicitly asked about the contrast between genetic modification and traditional methods of selective breeding. The speed and precision of new genetic techniques worried many, not least when placed in the context of what Macnaghten interprets as his respondents' desire to establish a relationship of respect with nature.[4] In this context, people become progressively more sensitive to their responsibilities for, and protection of, nature, including animals. Alongside that, people are skeptical of the capacity of companies or governments to exercise caution, and "spoke of the need for new institutional bodies that recognized the texture of current public responses, composed of people encompassing a broad spectrum of lay public opinion" (Macnaghten, 2001, p. 29).

Overwhelmingly, participants showed a "reaction against the proposed technology as intrinsically a violation of nature and transgressive of so-called natural parameters" (p. 25)—what might be called the "yuk response." Concerns over the speed and accuracy of crossing genetic boundaries and the possibly disastrous consequences that might flow from these are folded into a deeply negative response to the mixing up of categories—different species—normally kept separate. As Mary Douglas (1966) argued many years ago, such impurity—matter out of place (in this case, the gene of one species placed in the genome of another)—provokes a profound sense of unease. Although these categories are culturally derived, they are the constructs through which we grasp the world, so when they are "illegitimately" combined, we become disoriented—our whole relationship to nature is under threat.[5]

There is another aspect to this uneasiness about transgenic animals, a fear of the consequence that nature might "bite back." That is, people are worried that genetically manufactured organisms may not be as predictable as scientists seem to suggest. Science is a messy and unpredictable practice, as several sociologists of science have emphasized (e.g., Collins, 1985; Lynch, 1985; Knorr-Cetina, 1999). Yet there is a pervasive image of science—one often promoted by scientists themselves—that it entails a rather linear process of discovery (and innovation).[6] In the case of transgenic animals this translates into the deliberate making-to-order of various species. Laboratory animals will not only be available simply "off the rack" as they are now (as we saw in chapter 2, strains of animals with particular characteristics are readily available from catalogs). Rather, we have entered an era in which some species can, it appears, be "redesigned" genetically so that certain characteristics desirable for a given biomedical purpose (as model or product) will be expressed; and scientists will now be able to have made-to-order animals with wholly novel qualities directly suited to particular experimental programs or systems. Whether this "technoscientific bespoking" happens or not is irrelevant here; what genetic engineering and cloning imply to many is a view of animals that reduces them to their genetic makeup and ignores unpredictability (see Ho, 1999).

Moreover, technoscientific bespeaking, in giving the impression that animals are wholly determinable—that is, wholly knowable—dramatically reduces the complex cultural role that animals have played in Western societies. Many authors have traced (e.g., Thomas, 1984; Baker, 1993) the densely ambiguous and contradictory symbolism that attaches to animals; this symbolism has functioned in all manner of ways in Western societies. Animals have served in our thinking about morality, truth, divinity, identity, risk—the list is endless. Ironically, it is, Michael (2001) suggests, the very unknowability of animals that allows them to play such a rich and multifarious role in our meaning-making systems. In genetically essentializing animals—that is, by ostensibly removing their otherness and unknowability—their role as symbolic resources becomes, perhaps catastrophically, diminished. As a result, how we understand ourselves, the animals around us, and the world at large becomes further constrained. The cultural shift born of this symbolic diminution also contributes to the sense of public unease about genetic engineering of animals (Schroten, 1997).

As we noted, people express more concerns over some species than others—primates and dogs are particularly well-regarded. Arguably, this focus on "established" species underpins the unease we detected with the transition of the lab animal from "model and strain" to "hybrid and product." If human capacity to intervene in the composition of lab animals has become qualitatively enhanced with the rise of new genetic technologies, this has led to profound concerns about the transgression of natural boundaries, both around the naturalness of species themselves and the very process of speciation, that is, how species naturally emerge through evolution. These concerns in turn assume particular forms of reproduction—namely, genetic material flows "nuptually," through sexual reproduction amongst members of the same species.

Yet the widespread cultural assumption that species are pure and separate has not gone unchallenged. Biologically, all of us are composite—guts are full of useful bacteria, cells of all higher organisms could not function without mitochondria which once were, in all likelihood, symbionts—possibly from bacteria (Margulis, 1993; Haraway, 1997). Haraway points out that species have never been pure—there is a "transversal communication" in which genetic material moves between even phylogenetically widely separated species, as well as the more obvious movement from one generation to the next. If we accept this impurity as a constitutive feature of nature, then transgenic animals such as the oncomouse—a mouse genetically altered to grow cancers—are not simply nature going about its usual business. Rather, they stand for much more than that, coming to embody some of our deepest fears about crossing species boundaries.

Our unease about the impurity of such figures is, suggests Haraway, ironic given that we now reside in an era of the mixed: the oncomouse is mixed in a multitude of ways—it is both biology and culture, commercial and academic, it is full of uncertainties and certainties. Such mixings represent both creativity and exploitation and render hybrid entities, such as oncomouse or transgenic donor pigs, profoundly ambiguous. These are at once a product of nature's creativity teased out by human creativity, and objects through which we understand nature. Whatever we do, we cannot separate ourselves from these hybrid entities—we are entwined culturally with them in complex and ambiguous ways, because nowadays it is through them that we come to grasp the natural world. As Myerson (2000) frames it, "we can acknowledge our kinship with these new possibilities, either as victims or as heroes" (p. 73). Perhaps it is such "mature ambivalence" about nature and science that is tapped by studies of public responses in which laypeople incorporate the contingencies of science, culture, and politics together. It is these complexities, in turn, which are embodied in the laboratory animal as both "model and strain" and "hybrid and product."

Making and Unmaking Animals and People

In this chapter, we have moved through various analyses of "what the public thinks." These range from surveys of public views of animal research, which assume the public to be deficient repositories of knowledge, to studies that argue that publics can be seen as sophisticated calculators of the costs and benefits associated with animal experimentation. These are, however, models of the public that reflect, and contribute to, traditional top-down relations between science and the public. "Facts" are key here: negatively, laypeople are assumed not to possess enough facts, while positively, they are seen to be able to draw rational conclusions once they have the facts. But in both cases, "facts" are uncertain. We have argued that people can be acutely aware of these uncertainties and the relations of trust that are implied by them. What this points to is that "the public" cannot easily be categorized as merely ignorant or ill-informed, as several spokespeople for research claimed.

Moreover, if progress in natural science is by no means as linear as scientists sometimes portray, neither are social scientific techniques themselves innocent. They make assumptions about, and generate knowledge of, people that in turn can feed into laypeople's self-understandings and experts' assessments of the public, and thus into arguments over animal experimentation. Indeed, Irwin and Michael (2003)[7] have argued that techniques such as surveys, given their institutional setting and widespread dissemination in the media, serve to establish particular models of citizenship and decision-making (e.g., how to act politically and responsibly). To the extent that these are taken up by laypeople, the very category of "lay people" is remade.

Yet these survey techniques hardly seem able to accommodate the multifarious and diverse concerns people have when it comes to animal experimentation. This complexity is redoubled when laypeople start to unravel the implications of the new genetics, whereby the laboratory animal is simultaneously many things. How does a questionnaire, for example, deal with the symbolic uncertainties thrown up by "technoscientific bespoking"?

For many authors working in the field of public understanding or engagement with science,[8] broadening democratic procedures to include the voices of lay publics is the best way of dealing with these uncertainties. If we return to the idea of "core sets," this means opening up the core set to a range of constituencies extending well beyond the expert (whether these be experts in science, ethics, or policy). Yet extending scientific decision-making into a wider domain is not unproblematic: it meets institutional resistance, for a start (see Elam and Bertilsson, 2001). It is also a problem in that it posits citizens as actors separate from the science who can be drawn in to comment on it. But what the studies of people's attitudes to, and perceptions of, scientific controversy have shown is that "the public" cannot so easily be separated; rather, "the public" emerges out of complex and dynamic cultural processes—which also involve scientists, governments, media, and animals. It is an idea (or ideas) that is constantly being made and unmade.

And of course, ideas of "publics" also emerge through the social scientific techniques used to study them, whether that is the very different approach of the survey or the focus group. The overarching point, which we hope to have illustrated in this chapter, is that our social scientific understanding of the public's views of animal experimentation cannot be divorced from the methodologies employed. As the laboratory animal is made and unmade, so too is the identity of the lay public.

That, perhaps, is what lies at the heart of the perennial controversy over animal experiments. It is, of course, about much more besides, and particularly it is about what actually happens to laboratory animals. But it is also, profoundly, about an ongoing battle for hearts and minds. We have explored in this part some of the ways in which researchers and their spokespeople engage with "publics" outside science; some, such as anti-vivisectionist organizations, are not likely to be persuaded and are likely to be vilified. Others are seen as more rational and are appealed to as such. But the common image conjured up by researchers is that the public is simply ignorant. As we have seen in this chapter, public responses cannot be so easily categorized as rational/irrational, knowledgeable/ignorant—they are much more complex than that and reflect layers of ambiguous responses both to animals and to science. Difficult though it would be, animal advocates and research advocates need to take these complex public anxieties to heart.

Conclusion: Who or What Is the Laboratory Animal?

In this book, we have considered the laboratory animal and its relationship to researchers from a number of diverse perspectives. The following list barely summarizes the versions of the lab animal we have encountered: It is an entity that is progressively developed, refined and marketed; it is an element in an experimental system that is both a unique individual and a standardized unit; in terms of physiology it can model the workings of human bodies; it is the object of ethical concern and political rhetoric; it epitomizes messy biology, which can precipitate disgust or enthrall with its intricacies; it is a symbol of hope for patients, and a sign of exploitation; it is an entity to be grasped through complex evolutionary understandings and a potential companion; it is a mechanical object and a complex subject; protected by properly legislated regulation and the victim of biomedical hubris; it is the precipitator of more or less uncontrollable emotions and a topic to be calmly assessed in terms of costs and benefits.

It is all of these and much besides. Within these broad characterizations lurk also identities—not least the identities of the scientists whom we have studied as they go about articulating and demarcating—that is, "performing" or "enacting" in the sense used by social theorists (for an excellent recent discussion see Mol, 2002)—their various relationships. These include relationships to laboratory animals themselves, to technicians and other scientists, to the processes of knowledge-making and ethical propriety, to rationality and emotionality, to regulators and scientific spokespersons, to patients and patient groups, to publics of various sorts, to animal welfare organizations and anti-vivisection groups.

Of course, among this morass of accounts, entities, relations, performances, and identities there are patterns to be discerned. Scientists are caught on the horns of various dilemmas[9] or dichotomies in which the need to address divergent constituencies—colleagues, publics, regulators, patients—expose fissures in any simple idea of a "scientific identity." For instance, making a transition in scientific training from seeing animals as naturalistic to animals as experimental objects does not mean that scientists can ignore norms associated with the naturalistic animal. On the contrary, scientists must continue to demonstrate their ethical concerns for animals.

Similarly, in regard to regulatory frameworks, scientists perceive the bureaucracy both negatively and positively. On the one hand, submitting to the regulations is a chore that is often seen to detract from the real and urgent scientific work that needs to be done. Regulation is seen as an imposition—an "other"—an unwelcome dimension of contemporary scientific practice and, indeed, identity. On the other hand, regulation is seen to "protect" scientists—it is an independent and politically grounded form of checking on scientists. Scientists thus differentiate themselves from, and align themselves to, the relevant regulatory bodies. Such identification and dis-identification have particular rhetorical effects. The former abstracts and privileges scientific procedures and processes; the latter embeds science within overtly politically responsible institutions. This is a rhetoric in which cakes are both had and eaten: privilege and responsibility are enacted here.

Let us take another example. Although animals are routinely represented as standardized, as the testimony of technicians shows, such standardization must be produced through various informal procedures of handling. In other words, the production of generalizable facts, so tied up with standardization, rests on nonstandardizable modes of interaction with individual animals. Scientific identity here reflects not only the dichotomous identities of animals (standard/nonstandard) but also divisions of labor and skill within the scientific community (scientists/technicians).

A final example is the way that scientists cycle through a range of identifications with, or distancing from, the public. As we have seen, key to such processes are judgments of rationality or emotionality. Publics are often seen to be highly irrational—overly sentimental about animals as a result of their ignorance of animals' natural history or too easily swayed by spectacular images of what appears to be animal suffering. Moreover, such emotionality is thought irrationally to neglect the fact that animal research is concerned with developing treatments or medicines to benefit people. In the process of such comparisons, scientists present themselves as possessing both the knowledge and the rationality to judge the necessity of both particular animal experiments and animal experimentation in general. As we have seen, at the same time, they too display "irrational" and emotionally laden decisions—say, with regard to limited choice of species they are willing to experiment on, or when they appear to grossly sentimentalize human sufferers, or when they lump together and demonize very different animal welfare and rights groups. What such displays do is enact an identification with (particular) publics—which is hardly surprising, as scientists are, after all, still part of the public.

If we have established that scientists enact shifting and multiple identities, these emerge in relation to the multiple identities of many other actors in this controversial domain—most obviously, perhaps, animals and publics. Needless to say, we doubt that we have done full justice to the complexity of "animals," "scientists," or "publics," nor, more importantly, to the interrelations between these. What we can say is that, however complex, it is the sustained study of these interrelations that might in the future better illuminate this controversy. And in an enduring controversy characterized by so many accusations of right and wrong, a recognition of complexity would be a good place to start. That in itself will not change the fact of animal use in invasive experiments, but it might provide a basis for dialogue that has been so sadly lacking.

Notes

Introduction

1. Quotations from websites: the first is from www.rds online.org.uk/pages/ home.asp and the second from www.animalaid.org.uk/viv/index.htm (accessed November 2005).

2. One theory, for example, is that the growth in animal rights and environmentalism is related to an emerging ontological insecurity as postmodernity progresses (Franklin et al., 2001). Animal rights might be one facet of a form of misanthropy that is "explicitly tied to a perceived crisis of morality and disorder in late modernity" (Franklin, 1999:197).

3. First, Arluke carried out a large ethnographic study of a number of U.S. labs (mostly in the Northeast), doing participating observation and interviews of personnel, particularly research scientists and animal caretakers/technicians. In addition, he carried out an analysis of images of laboratory animals, and how they were represented in specialist journals. He also studied (with Dorian Solot and Fredric Hafferty) the educational use of animals in biology classes and medical school, and analyzed (with Julian Groves) the pro-science movement. Secondly, Birke and Michael carried out a smaller project in the U.K., based on interviews, mainly with research scientists and some caretakers; this focused on the changing climate of animal research a few years after the Animals (Scientific Procedures) Act came into force in Britain in 1986. They have also done analyses of the language used to describe scientific procedures, both in specialist journals and in media representations. Birke has done a study of how lab rats are represented in journals, broadly similar to Arluke's. Finally, Birke and Michael have done several studies of how the public understands science (for example, in relation to xenotransplantation—the use of animals to provide organs for transplantation). We should emphasize here that these studies use a qualitative approach, which means that we quote extensively from interviews and from field notes—a sociological approach which may seem anecdotal to readers more familiar with quantitative studies. The published

journal articles based on this research can be found in the list of references at the end of the book.

4. This line of inquiry is often pilloried by scientists, who sometimes refuse to accept the idea that they might create a "fact" themselves—that the idea of a "fact" is socially constructed. Why does a plane stay in the air, they might argue, if aerodynamics is simply a social construct? The details of these debates are largely irrelevant here. The point, however, is that sociologists of science have mapped out the ways in which social interests have become part of the process of making knowledge. In the case of animals, for instance, widespread cultural beliefs about animals versus humans have bearing on the way that science generates knowledge about animals.

5. Defenders of animal use dislike this word, insisting that much research using animals does not involve any kind of cutting, and if it does, the animal is fully anesthetized. We use it occasionally in this book because it is in common usage, usually to indicate specifically physiological or surgical research.

6. Even after anesthesia had been introduced, scientists did not seem to use it all the time (see Rupke, 1987)—even into the twentieth century. Use of anesthesia is now required by law in the United States and Europe. Proportionately fewer experiments, moreover, now involve surgery, but the number of experiments conducted has grown massively since those early protests. Today, animals are used in millions not only for basic research but also to fulfil legal requirements that new products are tested for safety. And new techniques, such as genetic manipulation, create new demands for animal use.

7. Direct government intervention also helped to develop this argument, particularly in the U.S., where a Federal Bureau of Investigation inquiry was launched, around 1918, in response to anti-vaccination campaigns by anti-vivisectionist organizations. Records of this investigation are held in the National Archives, in Washington, DC: FBI files.

8. We say "many" here because anti-vivisectionist literature rarely addresses the whole of the animal kingdom. In general, their publicity material emphasizes mammals and birds—animals with much more impact than, say, cockroaches or nematodes. This is not, we should add, that anti-vivisectionist organizations believe that certain animals could be excluded, simply that their literature tends not to mention them.

9. Utilitarianism is the attempt to balance the greater good against the costs or sufferings. Thus, in animal experiments, the prevailing belief is that the greater good (potential medical benefits for many people) must outweigh the suffering (of a limited number of animals). It is not, of course, an easy balance to estimate.

10. Critics argue that this process of renegotiation ultimately permits most potential experiments. Whether that is the case or not, what the licensing system does not do is any auditing of experiments after they have taken place. There is no clear post hoc system for evaluation of experiments in terms of their potential benefits.

11. We cannot know what non-human animal identities mean to them in the way that we can ask other humans linguistically to talk about their sense of self-identity. The animal identities we describe are necessarily products of human constructions of animals. The point, however, is that the non-human participants are also important in the processes of identity formation.

12. We chose these categories of lab workers because they are, arguably, most directly involved in the day-to-day lab work. There are, of course, others who are involved in research, whom we did not interview, such as inspectors (for the U.S. Department of Agriculture, which oversees the Animal Welfare Act, or for the British Home Office). Both the U.S. and the U.K. now demand that labs have specified veterinarians; for extensive discussion of their role in research and the care of lab animals, see Carbone (2004).

13. A framework we find useful for looking at these multiple processes of creating identities is actor network theory, or ANT (see Latour, 1987; Michael, 1996). This has been developed particularly in connection with the sociology of scientific knowledge. Influenced by Foucault's mapping of how power operates through social and historical networks, ANT seeks to map out networks of interests and influence in scientific practices. Building on social constructionism, which focuses on how an identity or role might be constructed as the person engages with the social world (to produce gender identities, for instance), ANT seeks to interrogate the wider social and technological context, following all the different actors, both human and nonhuman (including machines, animals, institutions and so forth). In doing so, it seeks to not give priority to any one actor or set of processes.

Chapter I

1. In her history of the development of purpose-bred mice in American genetics research, Karen Rader notes how, at the Jackson Laboratory in Bar Harbor, Maine, Clarence Cook Little was able to get mice established as research organisms partly because these animals did not readily challenge cultural antipathy toward animal experiments (Rader, 2004:29).

2. Harriet Ritvo (1987) has charted the selective breeding of certain kinds of animals in the Victorian era, and its relation to social class—though why rat and mouse breeding were developed in working-class communities remains unclear. In Britain, pigeons and whippets have also been traditionally bred in working-class communities, particularly among miners.

3. The Jackson Laboratory has been developing rat and mouse strains for research since the early twentieth century. See the JAX website, at http:/www. jax.org, for listings of currently available mouse strains and chromosome mappings.

4. These are produced by Caesarian section, and the uterus passed into a sterile environment and the foetuses fostered onto an already sterile mother. The biological mother's death is not typically noted in descriptions of how such animals are derived.

Chapter II

1. James Tiptree Jr. was the pseudonym of psychologist and science fiction writer Alice Sheldon.

2. This is not to say that this is the best or only way of generating knowledge, simply that historically, animals have indeed been used.

3. Cage sizes are now subject to legislative requirements, specifying minimum sizes and maximum stocking densities; in the U.K., see guidelines issued by the Home Office. Minimum cage size is, not surprisingly, much disputed; see Carbone (2004). For some species, there are also requirements that the animals are taken out of their cages for exercise. There is, however, now much more interest in methods of improving both the quantity and quality of internal spaces inside laboratory cages by, for example, providing "environmental enrichment," such as toys or things to hide inside.

4. Animal houses are, in effect, a kind of panopticon, incorporating power and control over the animal occupants (and the human ones); see Foucault (1979) and elaboration of this idea with respect to zoo animals in Acampora (2005).

5. This shift within science mirrors a shift within the wider culture, in which animal suffering at human hands has become less and less visible. See Burt's (2001) discussion of visualization technologies and animal representation.

6. Technicians do, of course, have their own professional organizations, training and guidelines, while information is exchanged through conferences and papers in specialist journals such as *Laboratory Animal*. What we emphasize here is that much of the knowledge about the actual handling of animals cannot easily be quantified or written down.

7. In a survey of laboratories carrying out mandatory toxicological testing, Hubrecht (1995) noted how variable reports of animal handling and enrichment were, with different labs reporting quite different amounts of animal interaction.

8. The local accomplishment of standardization is something that pervades technoscientific societies—even "standard" measures such as the volt have to be constantly calibrated. On the whole, people take such units for granted. Animals, however, are likely to be a different proposition, because of the uneasy tension between the standardized animal of science and the naturalized, individualized, creature we might encounter at home.

9. One reason may be the barriers of professionalization—how much do clinical researchers consult animal-based journals? Carbone (1994:ch. 6) discusses this point with respect to veterinary medicine.

10. They assessed evidence from evolution and genetics, and from several areas of biomedical research, before concluding that, in practice, animal models did not provide good bases for understanding human disease. However, they could sometimes have heuristic value in provoking research in humans (see also Shapiro, 2002).

11. Because of this, these authors argued for prospective registering of animal experiments, so that links between animal studies and clinical research could be better established even if the correlation is negative (Roberts et al., 2002). This could, in theory, be done in the U.K., for instance, through existing legislation governing animal use in experiments (in Britain, all such experiments must be licensed beforehand).

12. Also see Shapiro (2002), and Plous (1992) in relation to animal models in psychology.

13. There has been a great deal of interest in recent years, from within the scientific community itself, in what is now called welfare science, which aims to assess the welfare needs of animals. It is ironic that this branch of science, using scientific methods to address the well-being of animals, has thus provided grounds for undermining belief in scientific methods—that is, the suggestion that how we keep animals in cages may affect their welfare, and hence the replicability or validity of the experiment itself.

14. For example, discussing the mixed results obtained from using transgenic animals, Wolozin (2001) went on to conclude that information gained from transgenics would lead to profound insights. Posted on the website for the Alzheimer Research Forum, web search May 2004.

15. Michael Hayre, director of the laboratory animal research center at Rockefeller University, cited in Ahern, 1995.

16. See Taconic Farms website, searched May 2004.

17. This is not to say that welfare is ignored, as there are growing demands for laboratories in North America and Europe to adhere to certain minimum standards of welfare. But the point here is that the more that individual variation is ignored, the easier it is to ignore individual welfare needs.

Chapter III

1. One consequence is that students often fail to grasp the idea that day-to-day science is messy; it is very rarely the series of perfectly executed experiments that seem to emanate from the research report.

2. This study analyzed 149 scientific papers, from eight journals, published in 1990–1991, focussing explicitly on the phrasing used in methods sections and what was included or excluded.

3. See Birke and Smith (1995:31). In this analysis of scientific papers, Birke and Smith deliberately did not cite references for each example given, precisely because they were examples of a general trend in scientific reporting.

4. For example, advertisements in *Science* in 2000–2001 included one for BD Biosciences, featuring a cheetah, with the headline "Counting elispots at high speed requires a new level of sophistication," while Roche Diagnostics used a poison dart frog with the heading "Maximise the Possibilities in Protein Expression." As the color of the frog depends inevitably on protein expression, its intensity is intended to convey a sense of greater expression.

5. Based on analysis of 90 advertisements over a ten-year period. See Arluke (1994). We have also examined more recent advertisements in laboratory animal publications in writing this chapter.

6. Ad for Charles River Laboratories. See Arluke (1994).

7. For example, an ad in *Science* for DirectGenomics depicted a shopping bag with a rat, a mouse, an ear of corn, grapes and a cow emerging from the bag under the caption, "Your one-stop source for homology driven genomic services."

8. Seventy-seven percent in Arluke's study. This use of animal images without a context remains common on websites for animal breeders.

9. Ad for Marshall, Europe. Marshall Farms and their affiliate, Butler Farms, produce purpose-bred mongrels and hounds for laboratory research, as well as ferrets and a range of pet-care products. Given the implications of purity and health in the ad, it is ironic that they have also been the focus of animal rights protests alleging dirty conditions and insufficient environmental enrichment for the dogs in Marshall's U.S. sites (information from website for People for the Ethical Treatment of Animals, accessed April, 2004).

10. Ad for Hazleton Research Animals, reproduced in Arluke, 1994.

11. Ad for Charles River Laboratories, 1997.

12. Ad for GeneSpring, *Science* 2001.

13. Ad for Covance (formerly HRP Inc)., 2001.

14. Of course, this is not always a dilemma, and many cultures within and beyond the West happily inhabit both conventions simultaneously, or rather, inhabit something like hybrid conventions.

15. Significantly, many of the interviewees in one of our research studies emphasized that animal welfare was respected in their own institution, even if not elsewhere. This tension between seeing the animal as agential versus tool also maps onto the polarization between animal rights (generally abolitionist) and animal protection (generally reformist, and advocating welfare improvements within labs). See Francione, 1996.

Chapter IV

1. See Armstrong (2004) for an analysis of changing "structures of feeling" toward animals as evident in literature. He uses the example of Melville's novel *Moby-Dick*. Similarly, Fudge (2000) analyzes the human-nonhuman relationship in the rapidly changing culture of early modernity.

2. This is not to say that the stance of "objectivity" is either truly attainable or devoid of emotions. On the contrary, rational objectivity is itself a product of, and influences, particular ways of constructing emotions; see Williams (2001) for further discussion of emotions and social theory.

3. Dr. Tannetje Koning, veterinarian in Zeewolde, The Netherlands, cited on interNICHE website, "Criticisms of harmful use," accessed May, 2005.

4. Fetal pigs are a by-product of the meat-packing industry; fetuses from pregnant sows are sold to biological supply houses for dissection purposes. However, none of the students appeared to know this.

5. Downie and Meadows (1995) reported that cumulative examination results of 308 undergraduate biology students who studied model rats were the same as those of 2,605 students who performed rat dissections, suggesting that opting out of dissection did not impair learning.

6. Requiring students to use live animals is much less common in the U.K.

7. One of the authors of this book (LB) had a similar experience in her training in biology when confronted with frogs killed at the start of a neurophysiology practical. Despite her utter revulsion at the whole thing, she well recalls overpowering awe at the sight of functioning frogs' bodies—a contradiction indeed (see Birke, 1994).

8. Using animals from the pound is sometimes permitted in the United States, but is not allowed under any circumstances under British law.

9. It is indeed a double irony when the animal flesh in question is the heart—culturally so strongly identified with the emotions.

10. Student comment posted to website, http://www.cfhs.ca/humaneeducator/ HE2001-1/he01-1p4a.htm (search July 21, 2004).

11. Most students are not, after all, headed for those careers relying on surgical skills. For the majority of students, computer simulations may well teach them all the anatomy they need to know.

12. Although, as Harré points out, "bodily feeling is often the somatic expression to oneself of the taking of a moral standpoint" (1991:143). Harré, writing about anger, argues that there is no such thing as emotion per se, only ways of acting emotionally.

13. For further discussion on "learning how to see" in science education, see Keller (1996) and Birke (1999).

Chapter V

1. This is an example of what Nelson (1993) calls epistemological communities. By this she means the way in which knowledge is created collectively, through debate and the creation of shared meaning, *within* different communities. Science is no exception here.

2. Whether this can ever be judged is debatable. To kill rats, a rodent guillotine is routinely used in labs and assumed to be relatively humane (because it is quick). Open to question is how humane it is, as Carbone (2004) discusses.

3. The speaker was referring to possible behavioral changes: because the fish continued to swim around, apparently normally, it was deduced not to be in pain. It is, however, well-established that receptor molecules for endorphins—the body's inbuilt painkillers—exist throughout the vertebrates, and in at least some invertebrates, too.

4. This is not to say that the anesthesia was always well-used, Phillips (1994) points out. Sometimes, insufficient time is allowed for full anesthesia to develop and animals undoubtedly suffer pain. In both the U.S. and the U.K., moreover, pain relief is required under current legislation.

5. The rabbits were supposed to be killed by exsanguination. So the technicians collected a large volume of blood from him over several days, and fed him vitamins and fluids to sustain him. The ruse worked.

6. See Irvine (2004) and Birke and Michael (1997) for discussion of human-animal relationships.

7. For discussion of the fissure between subjective and objective perspectives on animal minds and relationships, see Dutton and Williams (2004).

Chapter VI

1. In practice, very few prosecutions are ever brought under existing legislation. More commonly, infringements might result in denial of licenses or funding, or the establishment of public inquiries, depending on the political culture of the country involved.

2. The Animal Technicians Association (later to become the Institute of Animal Technology), for example, was founded in Britain in 1950. In the U.S., the Animal Care Panel was established also at this time, later to become the American Association for Laboratory Animal Science (*Institutional Animal Care and Use Guidebook*, 2nd edition, 2002: ARENA/OLAW).

3. See J. Brown, letter to the *Spectator*, 26 June, 1875. Not long after the act was passed, the scientific community began to mobilize, forming the Research Defence Society, which is still in existence. Also see discussion in Rupke, 1987.

4. Obviously, this does not stop bad or cruel research in itself; but it does help to provide a climate in which ethical concerns take greater priority than they once did.

5. For a summary of policies and legislation within the countries of the European Union, see the European Science Foundation website, http://www.esf.org/publication/115/ESPB15.pdf.

6. In the same year, the European Union adopted a resolution requiring member states to regulate animal experiments.

7. This is not to say that pain is necessarily well-controlled, just that the law stipulates that scientists attempt to do so.

8. See Carbone (2004) for discussion of the care and use distinction in U.S. legislation.

9. This is the situation at the federal level, and for federal granting agencies; there may be stricter regulations at the state level, which would take priority.

10. Housing conditions are inevitably criticized by those opposed to research, and are often considered inadequate. Many scientists would agree, and there are many attempts in labs to provide environmental enrichment for animals.

11. The interview was carried out five years after the 1986 legislation came into force, so it was relatively new and unfamiliar at the time of the interview.

12. P. Gerone, Tulane Regional Primate Center, quoted in *Chronicle of Higher Education*, Sept. 4, 1991: S. Burd, "Critical of US rules on animal rights, scientists prepare to comply."

13. These were people sampled in the category of Personal Licence Holders, who are statistically likely to be more junior (Purchase and Nedeva, 2001a).
14. Proportions vary somewhat in different groups of people sampled; see Purchase and Nedeva (2001a) for details.
15. The British BBSRC website specifically refers to the need for researchers to develop a "culture of care" in relation to any funded research.

Note to Part III Introduction

1. In addition to the interviews with scientists, technicians, and administrators from which we have quoted in previous chapters, we also draw here on fieldwork notes. These come particularly from attendance at pro-research workshops and conferences (N=10) in the United States. This gave us the opportunity to talk informally with pro-research advocates and to observe discussions among them. We learned of these events from our informants, personal contacts, and through advertisements posted on notice boards in institutions. At observed workshops and conferences, representatives from different organizations gave presentations about the benefits of animal research. In the workshops, trainee animal researchers learned not only the regulations governing the use of animals in research, but also about the animal rights movement and strategies for dealing with its proponents. Finally, we collected and analyzed literature issued by the pro-research movement. In some cases, we obtained these from the conferences that we attended. In others, we wrote to the organizations themselves, after learning about them from the conferences and workshops.

Chapter VII

1. We quote extensively from fieldnotes and personal observations here. Other sources are referenced in the text.
2. British pharmacologist William Paton referred to this as "the depreciation of man," by which an animal rights position diminishes humankind's achievements; see Paton (1993).
3. Anita Guerrini (2003) notes how little Harvey is criticized by anti-vivisectionists; rather, it is Descartes, and his belief that animals did not have souls and were automata, who is consistently vilified in the animal rights literature. As she points out, Harvey carried out a great many experiments on living animals—unlike Descartes, whose concern was largely philosophical. Brandon Reines, whose writing was mentioned by the spokesperson cited here, has written articles for the Medical Research Modernization Committee, a group generally hostile to animal research.
4. Needless to say, a similar patterning can be claimed to characterize the discourse of pro-research advocates. For example, Brown and Michael (2001) note how advocates of xenotransplantation could be accused of disguising their

real research agenda when they insist on their commonsense, and thus the public representativeness of their views (about the moral standing of pigs).

5. See Home Office 2004 Report on Animal Welfare: Human Rights—Protecting People from Animal Rights Extremists.

6. Shaoni Bhattacharya: "Scientists demand law against animal rights extremism," April 2004, NewScientist.com, news service.

7. This is, of course, a claim highly contested by anti-vivisection organizations, which point out that it relies on an internalist view of medical history as progress. Their emphasis—similar to that made by many critical historians of science and medicine—is that diseases have often been brought under control at least partly by factors outside biomedicine itself—through improvements in sanitation, for example.

8. In the late nineteenth century, the medical community also sought to win public support by exaggerating the efficacy of treatments that came out of animal research. According to Turner (1980), claims were made that treatments had cut the mortality rate of diphtheria by as much as 40% when, in actuality, figures were closer to 10%.

9. This is not to say that all patients draw on this discourse. Some patients are well aware that biomedical spokespersons can overstate the importance of breakthrough and the imminence of any treatment that follows from such breakthrough (Michael and Brown, 2003a).

10. Paul (1995) similarly noted how the rhetoric of scientists and animal rights activists often took similar form.

Chapter VIII

1. Media representations of science illustrate the change in public trust toward science that is reflected in this statement. In the 1950s, for example, it was possible to portray dogs photographically as participants in research programs (see Carbone, 2004:80) in ways that became much more difficult later. LaFollette (1990) similarly tracks media representations of science, pointing out the loss of trust as science became more entwined through the twentieth century with national interests.

2. These are clearly not mutually exclusive; respondents might use more than one category in speech. We want to emphasize, moreover, that this typology is not unique to people working with animals; it would apply just as well to others working in stigmatized professions.

3. That risk was mentioned much less often in the British interviews, possibly because they were done through *New Scientist* magazine, and one of the interviewers is herself a biologist. Several commented that that situation felt less risky than talking to more varied media.

4. Scientists are not, of course, unique in using such rhetorical strategies; it is an obvious feature of political processes in general. But it is a crucial part of creating the *authority* of science.

5. The classic example of actor network theory is Callon's study of the scallop fishermen of St. Brieuc Bay, in northern France, and the biologists who sought to persuade them of the need for restocking (Callon, 1986). Callon describes the way that the researchers attempted to enroll the fishermen and the scallops, in different ways, in order to support their claims, thus creating potential networks. The scallops and fishermen, of course, also had other interests.

6. In addition, in the context of criticisms of the animal rights lobby's propaganda, there is a tacit criticism of the general public as prone to spectacle. Michael and Brown (2005) have explored this representation of the public (along with the charge of fickleness) in relation to the xenotransplantation issue.

7. Scientists are, of course, likewise stereotyping the lay public in such characterization.

8. Of course, the contrast between rationality and emotion is a highly spurious one, not least because we can be passionate about rationality, and rational about emotion (see, for example, Barbalet, 2001).

Chapter IX

1. There have been several approaches to promoting wider public consultation, such as "citizens' juries," which deliberate on a specific question and deliver a report. Some of these consultations are also tied in to qualitative research, such as those conducted by the Wellcome Trust. See, for example, their *Public Perspectives on Human Cloning* (London: The Wellcome Trust, 1998).

2. Xenotransplantation involves the transplantation of organs or tissues from one species into another, notably from non-human mammals into humans.

3. Other studies have also indicated considerable public distrust of scientists and of those institutions in charge of scientific research, particularly in relation to those areas of science raising ethical dilemmas. See, for example, comments from participants noted in the Wellcome Trust's report on cloning, note 1.

4. This fits with what Franklin (1999) calls risk reflexivity. According to Franklin, people have become increasingly aware that the "natural," from the environment to meat, is compromised by various forms of problematic intervention; because of this, the natural becomes "a permanent political fixture" (p. 59).

5. Of course, it is possible to provide a rather different interpretation of this reaction. As we saw, a key concern is that scientists are getting beyond themselves, they are "playing God." Here, ironically, the transgression is not between categories of nature and culture, but between those of the human and the divine.

6. Scientists themselves know, of course, how messy and unpredictable it is. The point here is that the public image of scientific practice and knowledge they help to perpetuate is one of a neatly tied linearity.

7. This suggestion follows arguments in the literature on governmentality (e.g., Dean, 1999).
8. See, for example, Funtowicz and Ravetz (1993); Joss (1999); Macnaghten (2001); Abelson et al. (2003).
9. These might be considered rhetorical dilemmas that reflect commonplace arguments and counter-arguments. See, for example, Billig et al. (1988).

References

Abbott, A. (2004). The Renaissance rat. *Nature* 428: 464–466, from Nature website, Science Update, April.

Abelson, J., Forest, P.-G., Eyles, J., Smith, P., Martin, E., & Gauvibin, F.-P. (2003). Deliberations about deliberation: Issues in the design and evaluation of public consultation processes. *Social Science and Medicine, 57*, 239–251.

Acampora, R. (2005). Zoos and eyes: Contesting captivity and seeking successor practices. *Society and Animals, 13*, 69–88.

Ahern, H. (1995). The rodent revolution. *The Scientist, 9(14)*, July 10.

Albury, D. and Schwartz, J. (1982). *Partial progress: The politics of science and technology.* London: Pluto Press.

Aldhous, P., Coghlan, A. and Copley, J. (1999). Let the people speak. *New Scientist, 22*, May, 12–14.

American Medical Association (1992). *A miracle at risk.* Chicago: AMA.

ARENA/OLAW (Applied Research Ethics National Association/Office of Laboratory Animal Welfare) (2002). *Institutional animal care and use committees guidebook.*

Arluke, A. (1988). Sacrificial symbolism in animal experimentation: Object or pet? *Anthrozoös, 2*, 97–116.

Arluke, A. (1990). Moral elevation in medical research. *Advances in Medical Sociology, 1*, 189–204.

Arluke, A. (1990). Uneasiness among animal technicians. *Laboratory Animal, 19*, 20–39.

Arluke, A. (1991). Going into the closet with science: Information control among experimenters. *Journal of Contemporary Ethnography, 20*, 306–330.

Arluke, A. (1992). Trapped in a guilt cage. *New Scientist, 134(1815)*, 33–35.

Arluke, A. (1994). We build a better beagle: Fantastic creatures in lab animal ads. *Qualitative Sociology, 17*, 143–158.

Arluke, A. (2004). The use of dogs in medical and veterinary training: Understanding and approaching student uneasiness. *Journal of Applied Animal Welfare Science, 7,* 193–204.

Arluke, A., & Hafferty, F. (1996). From apprehension to fascination with "dog lab": The use of absolutions by medical students. *Journal of Contemporary Ethnography, 25,* 201–225.

Arluke, A., & Sanders, C. (1996). *Regarding animals.* Philadelphia: Temple University Press.

Armstrong, P. (2004). Moby Dick and compassion. *Society and Animals, 12,* 19–38.

Bakan, D. (1968). *Disease, pain and sacrifice.* Chicago: University of Chicago Press.

Baker, S. (1993). *Picturing the beast: Animals, identity and representation.* Manchester: Manchester University Press.

Baker, S. (2000). *The postmodern animal.* London: Reaktion Books.

Barbalet, J. M. (2001). *Emotion, social theory and social structure.* Cambridge: Cambridge University Press.

Barr, G., & Herzog, H. (2000). Fetal pig: The high school dissection experience. *Society and Animals, 8,* 53–70.

Bazerman, C. (1988). *Shaping written knowledge: The genre and activity of the experimental article in science.* Madison: University of Wisconsin Press.

Beck, U. (1992). *The risk society.* London: Sage.

Bekoff, M. (2003). *Minding animals: Awareness, emotions, and heart.* New York: Oxford University Press.

Bendelow, G. and Williams, S. J. (Eds.) (1998). *Emotions in social life: Critical themes and contemprary issues.* London: Routledge.

Benton, T. (1993). *Unnatural relations.* London: Verso.

Best, J. (1987) Rhetoric in claims-making: Constructing the missing children problem. *Social Problems, 34,* 101–21.

Bhattacharya, S. (2004). Scientists demand law against animal rights extremism. April. NewScientist.com, news service.

Billig, M., Condor, S., Edwards, D., Gane, M., Middleton, D. & Radley, A. (1988). *Ideological dilemmas.* London: Sage.

Birke, L. (1994). *Feminism, animals and science: The naming of the shrew.* Buckingham: Open University Press.

Birke, L. (1999). *Feminism and the biological body.* Edinburgh: Edinburgh University Press.

Birke, L. (2003). Who—or what—are the rats (and mice) in the laboratory? *Society and Animals, 11,* 207–224.

Birke, L., Brown, N., & Michael, M. (1998). The heart of the matter: Animal bodies, ethics and species boundaries. *Society and Animals, 6(3),* 245–261.

Birke, L., & Michael, M. (1997). Hybrids, rights and their proliferation. *Animal Issues, 1(2),* 1–19.

Birke, L., & Smith, J. (1995). Animals in experimental reports: the rhetoric of science. *Society and Animals, 3,* 23–42.

Boyd Group (2001, March). Response to Home Office review of the Ethical Review Process. Retrieved from http://www.boyd-group.demon.co.uk.

Bradshaw, R. H. (2002). The ethical review process in the UK and Australia: The Australian experience of improved dialogue and communication. *Animal Welfare, 11*, 141–156.

Breakwell, G. (2002). Public attitudes to biotechnology with animals: A review. Retrieved from http://www.aebc.gov.uk/aebc/pdf/breakwell_literature_review.pdf.

British Association for the Advancement of Science (1993). *Animals and the Advancement of Science*. London: BAAS.

Brown, J. (1875). Letter. *The Spectator*, 26 June, 817.

Brown, N. (2000). Organizing/disorganizing the breakthrough motif: Dolly the cloned ewe meets Astrid the hybrid pig. In N. Brown, B. Rappert, & A. Webster (Eds.), *Contested futures* (pp. 87–108). Aldershot: Ashgate.

Brown, N. & Michael, M. (2001). Switching between science and culture in transpecies transplantation. *Science, Technology and Human Values, 26(1),* 3–22.

Bryld, M. (1998). The days of dogs and dolphins: Aesopian metaphors of Soviet science. In M. Bryld & E. Kulavig (Eds.), *Soviet civilization between past and present* (pp. 53–75). Odense, Denmark: Odense University Press.

Burd, S. (1991, Sept. 4). Critical of US rules on animal rights, scientists prepare to comply. *Chronicle of Higher Education.*

Burian, R. M. (1993). How the choice of experimental organism matters: Epistemological reflections on an aspect of biological practice. *Journal of the History of Biology, 26*, 351–367.

Burt, J. (2001). The illumination of the animal kingdom: The role of light and electricity in animal representation. *Society and Animals, 9*, 203–228.

Caine, N. G. (1992). Humans as predators: Observational studies and the risk of pseudohabituation. In H. Davis, & D. Balfour (Eds.), *The inevitable bond: Examining scientist-animal interactions.* Cambridge: Cambridge University Press.

Callon, M. (1986). Some elements in a sociology of translation: Domestication of the scallops and fishermen of St. Brieuc's Bay. In J. Law (Ed.), *Power, action and belief* (pp. 196–233). London: Routledge and Kegan Paul.

Carbone, L. (2004). *What animals want: Expertise and advocacy in laboratory animal welfare policy.* Oxford: Oxford University Press.

Cavalieri, P., & Kymlicka, W. (1996). Expanding the social contract. *Etica and Animali, 8*, 5–33.

Chesler, E. J., Wilson, S. G., Lariviere, W. R., Rodriguez-Zas, S. & Mogil, J. S. (2002). Influences of laboratory environment on behavior. *Nature Neuroscience, 5*, 1101–1102.

Clause, B. T. (1993). The Wistar rat as a right choice: Establishing mammalian standards and the ideal of a standardized animal. *Journal of the History of Biology, 26*, 329–349.

Collins, H. M. (1981). The place of the core set in modern science: social contingency with methodological propriety in science. *History of Science, 19,* 6–19.

Collins, H. M. (1985). *Changing order.* London: Sage.

Collins, H. M. (1988). Public experiments and displays of virtuosity. *Social Studies of Science, 18,* 725–748.

Collins, H. M., & Evans, R. (2002). The third wave of science studies: Studies of expertise and experience. *Social Studies of Science, 32(2),* 235–296.

Cunningham, P. F. (2000). Animals in psychology education and student choice. *Society and Animals, 8,* 191–212.

Davis, H., & Balfour, D. (Eds.) (1992). *The inevitable bond: Examining scientist-animal interactions.* Cambridge: Cambridge University Press.

Dean, M. (1999). *Governmentality: Power and rule in modern society.* London: Sage.

Denver, R., et al. (1988). Direct action for animal research. *Science 241,* 11.

Despret, V. (2004). The body we care for: Figures of Anthropo-zoo-genesis. *Body & Society, 10,* 111–134.

Dewsbury, D. A. (1992). Studies of rodent-human interactions in animal psychology. In H. Davis & D. Balfour (Eds.), *The inevitable bond: Examining scientist-animal interactions.* (pp. 27–43) Cambridge: Cambridge University Press.

Dickens, P. (1996). *Reconstructing nature: Alienation, emancipation and the division of labour.* London: Routledge.

Douglas, M. (1966). *Purity and danger.* London: Ark.

Downie, R., & Meadows, J. (1995). Experience with a dissection opt-out scheme in university level biology. *Journal of Biological Education, 29(3),* 187–194.

Driscoll, J. W. (1992). Attitudes toward animal use. *Anthrozoös, 5,* 32–39.

Dror, O. (1999). The affect of experiment: The turn to emotions in Anglo-American physiology. *Isis, 90,* 205–237.

Drury, M. (1991). Letter cited in *Research Defence Newsletter,* June, p. 3.

Durant, J. R. (1993). What is scientific literacy? In J. R. Durant & J. Gregory (Eds.), *Science and culture in Europe* (pp. 129–137). London: Science Museum.

Dutton, D., & Williams, C. (2004). A view from the bridge: Subjectivity, embodiment and animal minds. *Anthrozoös, 17,* 210–224.

Elam, M., & Bertilsson, T. M. (2001). Consuming, engaging and confronting science, The emerging dimensions of scientific citizenship. *European Journal of Social Theory, 2003(2),* 233–251.

Elston, M. A. (1987). Women and anti-vivisection in Victorian England, 1870–1900. In N. Rupke (Ed.), *Vivisection in historical perspective* (pp. 259–294). London: Routledge.

Eurobarometer 55.2 (2001). Europeans, science and technology. Retrieved from http://europa.eu.int/comm/research/press/2001/pr0612en-report.pdf.

Ewick, P., & Silbey, S. (2004). The architecture of authority: The place of law in the space of science. Retrieved from http://www.clarku.edu/activelearning/departments/sociology/ewick/ewick.cfm.

Fano, A. (1997). *Lethal laws: Animal testing, human health and environmental policy.* London: Zed Press.

Fleischmann, K. R. (2003). Frog and cyberfrog are friends: Dissection simulation and animal advocacy. *Society and Animals, 11,* 123–144.

Foster, H. L. (1980). The history of commercial production of laboratory rodents. *Laboratory Animal Science, 30,* 793–798.

Foster, J. (Ed.) (1997). *Valuing nature? Ethics, economics and environment.* London: Routledge.

Foucault, M. (1979). *Discipline and punish.* New York: Vintage Books.

Fox, M. (2004). Rat genome map shows similarity to humans. Retrieved April 28, 2004, from http://www.nlm.nih.gov/medlineplus/news/fullstory_16876.html.

Francione, G. L. (1996). *Rain without thunder: The ideology of the animal rights movement.* Philadelphia: Temple University Press.

Frank, R. G. (1994). Instruments, nerve action and the all-or-none principle. *Osiris, 9,* 208–235.

Franklin, A. (1999). *Animals and modern cultures: A sociology of human-animal relations in modernity.* London: Sage.

Franklin, A., Tranter, B., & White, R. (2001). Explaining support for animal rights: A comparison of two recent approaches to humans, nonhuman animals and postmodernity. *Society and Animals, 9,* 127–144.

Fudge, E. (2000). *Perceiving animals: Humans and beasts in early modern English culture.* Houndmills: Macmillan.

Fujimura, J. (1996). *Crafting Science: a sociohistory of the quest for the genetics of cancer.* Cambridge, Mass.: Harvard University Press.

Funtowicz, S. O., & Ravetz, J. (1993). Science for the post-normal age. *Futures, 25* (7): 735–755.

Galmark, L. (2000). Women antivivisectionists: The story of Lizzy Lind af Hageby and Leisa Schartau. *Animal Issues, 4,* 1–32.

Geison, G., & Laubichler, M.D. (2001). Reflections on the role of organismal and cultural variation in the history of the biological sciences. *Studies in History and Philosophy of Biological and Biomedical Sciences, 32,* 1–29.

Giddens, A. (1991). *Modernity and self-identity.* Cambridge: Polity Press.

Gieryn, T. (1983) Boundary-work and the demarcation of science from non-science: Strains and interests in the professional ideologies of scientists. *American Sociological Review, 48,* 781–795.

Ginsburg, B. E., & Hiestand, L. (1992). Humanity's "best friend": The origins of our inevitable bond with dogs. In H. Davis & D. Balfour (Eds.), *The inevitable bond: Examining scientist-animal interactions* (pp. 93–109). Cambridge: Cambridge University Press.

Gluck, J. P., & Kubacki, S. R. (1991). Animals in biomedical research: The undermining effect of the rhetoric of the besieged. *Ethics and Behavior, 1(3),* 157–73.

Goffman, E. (1959). *The presentation of self in everyday life.* Garden City, N.Y.: Doubleday.

Goffman, E. (1963). *Stigma: Notes on the management of spoiled identity.* Englewood Cliffs, N.J.: Prentice Hall.

Golding, C. (1990). *Rats: The new plague.* London: Weidenfeld and Nicolson.

Good, B. J. (1994). *Medicine, rationality and experience: An anthropological perspective.* Cambridge: Cambridge University Press.

Goodwin, F. (1992). Animal research, animal rights and public health. *Conquest, August (181),* 1–10.

Gross, A. G. (1990). *The Rhetoric of Science.* Cambridge, Mass.: Harvard University Press.

Groves, J. M. (1994). Are smelly animals happy animals? *Society and Animals, 2,* 125–44.

Guerrini, A. (1989). The ethics of animal experimentation in seventeenth-century England. *Journal of the History of Ideas, 50,* 391–408.

Guerrini, A. (2003). *Experimenting with humans and animals: From Galen to animal rights.* Baltimore: Johns Hopkins University Press.

Hacking, I. (1986). Making up people. In T.C. Heller, M. Sosna, & D.E. Wellberg (Eds.), *Reconstructing individualism* (pp. 222–236). Stanford, Calif.: Stanford University Press.

Hafferty, F. (1991). *Into the valley: Death and the socialization of medical students.* New Haven, Conn.: Yale University Press.

Hagelin, J., Carlsson, H. E., & Hau, J. (2002). An overview of surveys on how people view animal experimentation: Some factors that may influence the outcome. *Public Understanding of Science, 12,* 67–81.

Halpin, Z. T. (1989). Scientific objectivity and the concept of "the other." *Women's Studies International Forum, 12,* 285–94.

Haraway, D. (1997). *Modest_Witness@SecondMillennium: FemaleMan meets Onco-Mouse.* London: Routledge.

Harding, S. (1991). *Whose science, whose knowledge?* Buckingham: Open University Press.

Hardy, D. (1990). *America's new extremists: What you need to know about the animal rights movement.* Washington, D.C.: Washington Legal Foundation.

Harré, R. (Ed.) (1986). *The social construction of emotions.* Oxford: Blackwell.

Harré, R. (1991). *Physical being: A theory for a corporeal psychology.* Oxford, Blackwell. 142-3

Hendrickson, R. (1983). *More cunning than man: A complete history of the rat and its role in human civilization.* New York: Kensington Books.

Herzog, H. A. (1988). The moral status of mice. *American Psychologist, 43,* 473–474.

Herzog, H. A. (1993). The movement is my life: The psychology of animal rights activism. *Journal of Social Issues, 49,* 103–119.

Herzog, H., Rowan, A. N., & Kossow, D. (2001). Social attitudes and animals. In D.J. Salem & A.N. Rowan (Eds.), *The state of the animals* (pp. 55–69). Washington, DC: Humane Society Press.

Ho, M.-W. (1999). *Genetic engineering: Dream or nightmare.* Bath, UK: Gateway.

Holmes, F. L. (1993). The old martyr of science: The frog in experimental physiology. *Journal of the History of Biology, 26,* 311–328.

Home Office (1986). *Animals (Scientific Procedures) Act: Public general acts and general synod measures.* London: HMSO.

Home Office and the Department of Trade and Industry (2004, July). Animal welfare—Human rights: Protecting people from animal rights extremists. London: The Home Office Communication Directorate.

Horton, L. (1989). The enduring animal issue. *Journal of the National Cancer Institute, 81,* 736–43.

Hubbell, J. (1990). The "animal rights" war on medicine. *Reader's Digest, June.*

Hubrecht, R. (1995). *Housing husbandry and welfare provisions for animals used in toxicology studies: Results of a UK questionnaire on current practice (1994).* Wheathampstead, Hertsfordshire: Universities Federation for Animal Welfare.

Irvine, L. (2004). *If you tame me: Understanding our connection with animals.* Philadelphia: Temple University Press.

Irwin, A. (1995). *Citizen science: A study of people, expertise and sustainable development.* London: Routledge.

Irwin, A., & Michael, M. (2003). *Science, social theory and public knowledge.* Maidenhead, Berkshire: Open University Press/McGraw-Hill.

Jackall, R. (1988). *Moral mazes: The world of corporate managers.* New York: Oxford University Press.

Jasper, J., & Nelkin, D. (1992). *The animal rights crusade.* New York: The Free Press.

Joss, S. (1999). Introduction: Public participation and technology policy—and decision-making—ephemeral phenomenon or lasting change? *Science and Public Policy, 26(5),* 291–293.

Kaufman, S. R. (1993). Scientific problems with animal models. In A.N. Rowan & J.C. Weer (Eds.), *The value and utility of animals in research.* Report for Tufts Center for Animals and Public Policy, North Grafton, Mass.

Kean, H. (1998). *Animal rights: Political and social change in Britain since 1800.* London: Reaktion Books.

Keller, E. F. (1985). *Perspectives on gender and science.* New Haven, Conn.: Yale University Press.

Keller, E. F. (1996). The biological gaze. In G. Robertson, M. Mash, L. Tickner, J. Bird, B. Curtis, & T. Putnam (Eds.), *Future natural: Nature/science/culture* (pp. 107–121). London: Routledge.

Kevles, D. J., & Geison, G. L. (1995). The experimental life sciences in the twentieth century. *Osiris, 10,* 97–121.

Kleinman, D., & Kloppenburg, J., Jr. (1991). Aiming for the discursive high ground: Monsanto and the biotechnology controversy. *Sociological Forum, 6,* 427–447.

Knight, S., Vrij, A., Cherryman, J., & Nunkoosing, K. (2004). Attitudes towards animal use and belief in animal mind. *Anthrozoös, 17,* 43–62.

Knorr-Cetina, K. D. (1983). The ethnographic study of scientific work: Towards a constructivist interpretation of science. In K.D. Knorr-Cetina, & M. Mulkay (Eds.), *Science observed: Perspectives on the social studies of science* (pp. 115–140). London: Sage.

Knorr-Cetina, K. (1999). *Epistemic cultures: How the sciences make knowledge.* Cambridge, Mass.: Harvard University Press.

Kohler, R. (1994). *Lords of the fly: Drosophila genetics and the experimental life.* Chicago: University of Chicago Press.

Kohler, R. (2002). Labscapes: Naturalizing the laboratory. *History of Science, 40,* 473–501.

Koning, T. (2005). Quotation from InterNICHE website. Retrieved from http://www.interniche.org/critic.html.

Koski, E. (1988). The use of animals in research: Attitudes among research workers. In W. Kay, et al. (Eds.), *Euthanasia of the companion animal* (pp. 21–30). Philadelphia: Charles Press.

LaFollette, M. C. (1990). *Making science our own: Public images of science, 1910–1955.* Chicago: University of Chicago Press.

LaFollette, H., & Shanks., N. (1996). *Brute science: Dilemmas of animal experimentation.* London: Routledge.

Lane-Petter, W. (1952). Uniformity in laboratory animals. *Laboratory Practice, April, 30–34.*

Lansbury, C. (1985). *The old brown dog: Women, workers and vivisection in Edwardian England.* Madison: University of Wisconsin Press.

Lash, C., & Urry, J. (1987). *The end of organized capitalism.* Cambridge: Polity Press.

Latour, B. (1983). Give me a laboratory and I will raise the world. In: K. Knorr-Cetina & M. Mulkay (Eds.), *Science observed* (pp. 141–170). London: Sage.

Latour, B. (1987). *Science in action: How to follow engineers in society.* Milton Keynes: Open University Press.

Latour, B. (1990). Drawing things together. In M. Lynch, & S. Woolgar (Eds.), *Representations in scientific practice* (pp. 19–68). Cambridge, Mass.: MIT Press.

Latour, B. (1993). *We have never been modern.* Hemel Hempstead, UK: Harvester.

Latour, B., & Woolgar, S. (1979). *Laboratory life: The construction of scientific facts.* London: Sage.

Lederer, S. (1987). The controversy over animal experimentation in America, 1880–1914. In N. Rupke (Ed.), *Vivisection in historical perspective* (pp. 236–258). London: Routledge.

Lederer, S. (1992). Political animals: The shaping of biomedical research literature in twentieth-century America. *Isis, 83,* 61–79.

Lock, R. (1994). Dissection as an instructional technique in secondary science: Comment on Bowd. *Society and Animals, 2(1)*, 67–74.

Lockard, R. B. (1968). The albino rat: A defensible choice or a bad habit? *American Psychologist, 23*, 734–742.

Logan, C. A. (1999). The altered rationale for the choice of a standard animal in experimental psychology: Henry H. Donaldson, Adolf Meyer and "the" albino rat. *History of Psychology, 2*, 3–34.

Logan, C. A. (2001). Are Norway rats . . . things?: Diversity versus generality in the use of albino rats in experiments on development and sexuality. *Journal of the History of Biology, 34*, 287–314.

Logan, C. A. (2002). Before there were standards: The role of test animals in the production of empirical generality in physiology. *Journal of the History of Biology, 35*, 329–363.

Löwy, I. (2003). On guinea pigs, dogs and men: Anaphylaxis and the study of biological individuality, 1902–1939. *Studies in History and Philosophy of Biological and Biomedical Sciences, 34*, 399–423.

Lynch, M. (1985). *Art and artifact in laboratory science: A study of shop work and shop talk in a research laboratory.* London: Routledge.

Lynch, M. (1988). Sacrifice and the transformation of the animal body into a scientific object: Laboratory culture and ritual practice in the neurosciences. *Social Studies of Science, 18*, 265–289.

Lyotard, J.-F. (1984). *The postmodern condition: A report on knowledge.* Manchester: Manchester University Press.

Macnaghten, P. (2001). *Animal futures: Public attitudes and sensibilities toward animals and biotechnology in contemporary Britain.* Lancaster, UK: IEPPP, Lancaster University.

Maehle, A. H., & Trohler, U. (1987). Animal experimentation from antiquity to the end of the eighteenth century: Attitudes and arguments. In N. Rupke (Ed.), *Vivisection in historical perspective* (pp. 14–47). London: Routledge.

Manser, C. E. (1992). *The assessment of stress in laboratory animals.* Horsham, UK: RSPCA.

Margulis, L. (1993). (2nd ed.) *Symbiosis in cell evolution: Microbial communities in the Archean and Protozoic Eons.* New York: W.H. Freeman.

Maynard-Moody, S. (1992). The fetal research dispute. In D. Nelkin (Ed.), *Controversy: Politics of technical decisions* (pp. 3–25). London: Sage.

McCabe, K. (1990). Beyond cruelty. *The Washingtonian, 25(5)*, 73–195.

McOuat, G. (2001). From cutting nature at its joints to measuring it: new kinds and new kinds of people in biology. *Studies in the History and Philosophy of Science, 32*, 613–645.

Medical Research Council (1999). *Animals in medicine and science.* London: Medical Research Council.

Mercer, D., Inaba, M., Maekawa, F., Chen Ng, M., & Obata, H. (2002). Japanese attitudes toward xenotransplantation. *Public Understanding of Science, 11*, 347–362.

Michael, M. (1992). Lay discourses of science: Science-in-general, science-in-particular and self. *Science, Technology and Human Values, 17*, 313–333.

Michael, M. (1996). *Constructing identities*. London: Sage.

Michael, M. (2001). Technoscientific bespoking: Animals, publics and the new genetics. *New Genetics and Society, 20(3)*, 205–224.

Michael, M., & Birke, L. (1994). Enrolling the core set: The case of the animal experimentation controversy. *Social Studies of Science, 24*, 81–95.

Michael, M., & Birke, L. (1995). Animal experiments: Scientific uncertainty and public unease. *Science as Culture, 5*, 248–276.

Michael, M., & Brown, N. (2003a). Dystopias and dys-tropias: Futures and performativities in xenotransplantation. Paper presented at the Annual British Sociology Association Conference, University of York, April 11–13.

Michael, M., & Brown, N. (2003b). Xenotransplantation: Risk identities and the human/nonhuman interface. End of Award Report to the Economic and Social Research Council. Retrieved from http://www.regard.ac.uk/research_findings/L218252044/report.pdf.

Michael, M., & Brown, N. (2004). The meat of the matter: Grasping and judging xenotransplantation. *Public Understanding of Science, 13*, 379–397.

Michael, M., & Brown, N. (2005). Scientific citizenships: Self-representations of xenotransplantation's publics. *Science as Culture, 14(1)*, 38–57.

Michael, M., & Carter, S. (2001). The facts about fictions and vice versa: Process in the public understanding of science. *Science as Culture, 10(1)*, 5–32.

Midgley, M. (1992). *Science as salvation: A modern myth and its meaning*. London: Routledge.

Mol, A. 2002. *The body multiple. Ontology in medical practice*. Durham, N.C.: Duke University Press.

MORI (2002). The use of animals in medical research. Research conducted for Coalition for Medical Progress by Market and Opinion Research International. Retrieved from http://www.mori.com/polls/2002/cmp.shtml.

Mulkay, M. (1993) Rhetorics of hope and fear in the great embryo debate. *Social Studies of Science, 23*, 721–742.

Myers, G. (2002). Symbolic animals and the developing self. *Anthrozoös, 15*, 19–36.

Myerson, G. (2000). *Donna Haraway and GM Foods*. Cambridge: Icon.

Nelkin, D. (1992). Science, technology and political conflict: Analyzing the issues. In D. Nelkin (Ed.), *Controversy: Politics of technical decisions* (pp. ix–xxv). London: Sage.

Nelson, L. H. (1993). Epistemological communities. In L. Alcoff & E. Potter (Eds.), *Feminist epistemologies* (pp. 121–159). London: Routledge.

Nicoll, C. S. (1991). A physiologist's views on the animal rights/liberation movement. *The Physiologist, 34*, 303–315.

Noske, B. (1989). *Humans and other animals*. London: Pluto Press.

Nowotny, H., Scott, P., & Gibbons, M. (2001). *Re-thinking science: Knowledge and the public in an age of uncertainty*. Cambridge: Polity.

Orlans, F. B. (1993). *In the name of science: Issues in responsible animal experimentation*. New York: Sage.

Osborne, T., & Rose, N. (1999). Do the social sciences create phenomena? The example of public opinion research. *The British Journal of Sociology, 50(3)*, 367–396.

Paigen, K. (1995). A miracle enough: The power of mice. *Nature Medicine, 1*, 215–217.

Paris, S. (1992a). A rat is not a pig is not a boy! Letter to the Editor, *The Wall Street Journal*, 7 October.

Paris, S. (1992b). Lives aren't the same. Letter to the Editor, *Belleville News-Democrat*, 23 July.

Paris, S. (1992/93) Letter from the President. *Progress, 1(2)*, 1.

Paton, W. (1993). *Man and mouse: Animals in medical research*. Oxford: Oxford University Press.

Paul, E. (1995). Us and them: Scientists' and animal rights campaigners' views. *Society and Animals, 3*, 1–22.

Pennisi, E. (2004) Genomics. New sequence boosts rats' research appeal. *Science, 303*, 455–8.

Phillips, M. T. (1993). Savages, drunks and lab animals: The researcher's perception of pain. *Society and Animals, 1*, 61–81.

Phillips, M. T. (1994). Proper names and the social construction of biography: The negative case of laboratory animals. *Qualitative Sociology, 17*, 119–143.

Pickvance, S. (1976). "Life" in a biology lab. *Radical Science Journal, 4*, 11–28.

Pifer, L. K. (1996). Exploring the gender gap in young adults' attitudes about animal research. *Society and Animals, 4*, 37–52.

Plous, S. (1992). Attitudes toward the use of animals in psychological research and education: Results from a national survey of psychologists. *American Psychologist, 51*, 1167–1180.

Pound, P., Ebrahim, S., Sandercock, P., Bracken, M.B., & Roberts, I. (2004). Where is the evidence that animal research benefits humans? *British Medical Journal, 328* (28 February), 514–517.

Purchase, I. F. H. (1999). Ethical review of regulatory toxicology guidelines involving experiments on animals: The example of endocrine disruptors. *Toxicological Sciences, 52*, 141–147.

Purchase, I. F. H., & Nedeva, M. (2001). The impact of the introduction of the ethical review process for research using animals in the UK: Attitudes to alternatives among those working with experimental animals. *ATLA, 29*, 727–744.

Purchase, I. F. H., & Nedeva, M. (2002). The impact of the introduction of the ethical review process for research using animals in the UK: Implementation of policy. *Laboratory Animals, 36*, 68–85.

Purchase, I. F. H., & Nedeva, M. (2004). The impact of the introduction of the ethical review process for research using animals in the UK: Attitudes to training and monitoring by those working under the Animals (Scientific Procedures) Act 1986. *Animal Welfare, 13,* 205–210.

Rabinow, P. (1996). *Essays in the anthropology of reason.* Princeton, N.J.: Princeton University Press.

Rader, K. (1998). The mouse people: Murine genetics work at the Bussey Institution, 1909–1936. *Journal of the History of Biology, 31,* 327–354.

Rader, K. (1999). Of mice, medicine and genetics: C.C. Little's creation of the inbred laboratory mouse, 1909–1918. *Studies in the History and Philosophy of Science, 30,* 319–343.

Rader, K. (2004). *Making mice: Standardizing animals for American biomedical research 1900–1955.* Princeton, N.J.: Princeton University Press.

Radford, M. (2001). *Animal welfare law in Britain: Regulation and responsibility.* Oxford: Oxford University Press.

Regan, T. (1984). *The case for animal rights.* London: Routledge.

Richardson, R. (1987). *Death, dissection and the destitute.* London: Pelican.

Ritvo, H. (1987). *The animal estate: The English and other creatures in the Victorian Age.* Harmondsworth: Penguin.

Roberts, I., Kwan, I., Evans, P., & Haig, S. (2002). Does animal experimentation inform human healthcare?: Observations from a systematic review of international animal experiments on fluid resuscitation. *British Medical Journal, 324,* 474–476.

Rose, H. (1995). Learning from the new priesthood and the shrieking sisterhood: Debating the life sciences in Victorian England. In L. Birke & R. Hubbard (Eds.), *Reinventing Biology* (pp. 3–20). Bloomington, Ind.: Indiana University Press.

Rothschild, M. (1986). *Animals and man: The Romanes Lecture 1984–5.* Oxford: Clarendon Press.

Royal Society of London (1985). *The public understanding of science.* London: The Royal Society.

Rupke, N. (1987). Pro-vivisection in England in the early 1880s. In N. Rupke (Ed.), *Vivisection in historical perspective* (pp. 188–213). London: Routledge.

Rupke, N. (Ed.) (1987). *Vivisection in historical perspective.* London: Routledge.

Russell, N. (1986). *Like engend'ring like: Heredity and animal breeding in early modern England.* Cambridge: Cambridge University Press.

Russell, S. (1990). CFAAR "celebrates" world lab animal liberation week—with an iron lung! *CFAAR Newsletter, 3(1),* 1–3.

Ryder R. (1989). *Animal revolution: Changing attitudes toward speciesism.* Oxford: Blackwell.

Sales, G. (1988). Effects of environmental ultrasound on behaviour of laboratory rats. In *Laboratory animal welfare research: Rodents,* (pp. 17–23) Potters Bar, Universities Federation for Animal Welfare.

Sanders, C. R. (1999). *Understanding dogs: Living and working with canine companions*. Philadelphia: Temple University Press.

Schroten, E. (1997). Animal biotechnology, public perception and public policy from a moral point of view. Proceedings from an International Workshop on Transgenic Animals and Food Production, Stockholm, May 1997, 151–156. Retrieved from http://www.kslab.ksla.se/trannpdt.htm.

Serpell, J. (1986). *In the company of animals*. Oxford: Blackwell.

Shapin, S. (1988). Following scientists around. *Social Studies of Science, 18*, 533–550.

Shapin, S. (1991). Science and the public. In R.C. Olby et al. (Eds.), *Companion to the history of modern science* (pp. 990–1007). London: Routledge and Kegan Paul.

Shapiro, K. (1991). The psychology of dissection. *The Animals' Agenda, November*, 20–21.

Shapiro, K. (2002). A rodent for your thoughts: The social construction of animal models. In M. Henninger-Voss (Ed.), *Animals in human histories* (pp. 439–469). Rochester: University of Rochester Press.

Shepherd, P. (1996). *The others: How animals made us human*. Washington, D.C.: Island Press.

Sherwin, C. M. (2004). The influence of standard laboratory cages on rodents and the validity of data. *Animal Welfare, 13 (Supplement)*, "Science in the Service of Animal Welfare."

Singer, P. (1975). *Animal liberation*. London: Jonathan Cape.

Smart, K. (1993). Resourcing ambivalence: Dogbreeders, animals and the social studies of science. Ph.D. dissertation, University of Lancaster.

Smith, J., Birke, L., & Sadler, D. (1997). Reporting animal use in scientific papers. *Laboratory Animals, 31*, 312–317.

Solot, D., & Arluke, A. (1997). Learning the scientist's role: Animal dissection in middle school. *Journal of Contemporary Ethnography, 26*, 28–54.

Sperling, S. (1988). *Animal liberators: Research and morality*. Berkeley: University of California Press.

Staats, J. (1965). The laboratory mouse. In E.L. Green (Ed.), *Biology of the laboratory mouse, (2nd edition)*. New York: McGraw Hill.

Star, S. L., & Greisemer, J. R. (1989). Institutional ecology, "translations" and boundary objects: Amateurs and professionals in Berkeley's museum of vertebrate zoology, 1907–39. *Social Studies of Science, 19*, 387–430.

Tansey, E. M. (1994). Protection against dog distemper and dogs protection bills: The Medical Research Council and anti-vivisectionist protest, 1911–1933. *Medical History, 38*, 1–26.

Thomas, K. (1983). *Man and the natural world: Changing attitudes in England, 1500–1800*. London: Penguin.

Thomas, L. (1977). On the science and technology of medicine. In J. Knowles (Ed.), *Doing better and feeling worse* (pp. 35–36). New York: W. W. Norton and Company.

Tiptree, J. (1978). The psychologist who couldn't do awful things to rats. In J. Tiptree, *Star dongs of an old primate*. New York: Ballantine Books.

Todes, D. P. (1997). Pavlov's physiology factory. *Isis, 88*, 205–246.

Turner, J. (1980). *Reckoning with the beast: Animals, pain, and humanity in the Victorian mind*. Baltimore: Johns Hopkins University Press.

Turner, J. C. (1985). Social categorization and the self-concept: A social-cognitive theory of group behaviour. In E. J. Lawler (Ed.), *Advances in group processes: Theory and research* (vol. 2, pp. 47–74). Greenwich, CT: JAI.

Turney, J. (1998). *Frankenstein's footsteps: Science, genetics and popular culture*. New Haven, Conn.: Yale University Press.

Twining, H., Arluke, A., & Patronek, G. (2000). Managing the stigma of outlaw breeds: A case study of pit bull owners. *Society and Animals, 8*, 1–28.

Wahlsten, D., Metten, P., & Crabbe, J. C. (2003). A rating scale for wildness and ease of handling laboratory mice: Results for 21 inbred strains tested in two laboratories. *Genes, Brain & Behavior, 2*, 71–79.

Warren, C. (1974). *Identity and community in the gay world*. New York: John Wiley & Sons.

Wieder, D. L. (1980). Behavioristic operationalism and the life-world: Chimpanzees and the chimpanzee researchers in face-to-face interaction. *Sociological Inquiry, 50*, 75–103.

Williams, S. (2001). *Emotion and social theory*. London: Sage.

Wolozin, B. (2001). a-Synuclein and ß-amyloid: A linking of partners to model disease. Posted Oct. 8 to Alzheimer's Research Forum website (www.alzforum. org).

Wynne, B. E. (1991). Knowledges in context. *Science, Technology and Human Values, 16*, 111–121.

Wynne, B. E. (1992). Misunderstood misunderstanding: Social identities and public uptake of science. *Public Understanding of Science, 1*, 281–304.

Wynne, B. E. (1995). The public understanding of science. In S. Jasanoff, G.E. Markle, J.C. Peterson & T. Pinch (Eds.), *Handbook of science and technology studies* (pp. 361–388). Thousand Oaks, Calif.: Sage.

Wynne, B. E. (1996). May the sheep safely graze? A reflexive view of the expert-Lay divide. In S. Lash, B. Szerszynski & B. Wynne (Eds.), *Risk, environment and modernity* (pp. 44–83). London: Sage.

Wynne, B. (2003). Seasick on the third wave: Subverting the hegemony of propositionalism: A reply to Collins and Evans. *Social Studies of Science, 33*, 401–417.

Index